Studies in Systems, Decision and Control

Volume 543

The series "Studies in Systems, Decision and Control" (SSDC) covers both new developments and advances, as well as the state of the art, in the various areas of broadly perceived systems, decision making and control–quickly, up to date and with a high quality. The intent is to cover the theory, applications, and perspectives on the state of the art and future developments relevant to systems, decision making, control, complex processes and related areas, as embedded in the fields of engineering, computer science, physics, economics, social and life sciences, as well as the paradigms and methodologies behind them. The series contains monographs, textbooks, lecture notes and edited volumes in systems, decision making and control spanning the areas of Cyber-Physical Systems, Autonomous Systems, Sensor Networks, Control Systems, Energy Systems, Automotive Systems, Biological Systems, Vehicular Networking and Connected Vehicles, Aerospace Systems, Automation, Manufacturing, Smart Grids, Nonlinear Systems, Power Systems, Robotics, Social Systems, Economic Systems and other. Of particular value to both the contributors and the readership are the short publication timeframe and the world-wide distribution and exposure which enable both a wide and rapid dissemination of research output.

Indexed by SCOPUS, DBLP, WTI Frankfurt eG, zbMATH, SCImago.

All books published in the series are submitted for consideration in Web of Science.

Nguyen Hoang Phuong · Nguyen Thi Huyen Chau ·
Vladik Kreinovich
Editors

Machine Learning and Other Soft Computing Techniques: Biomedical and Related Applications

Springer

Editors
Nguyen Hoang Phuong
Artificial Intelligence Division
Faculty of Information Technology
Thang Long University
Hanoi, Vietnam

Nguyen Thi Huyen Chau
Artificial Intelligence Division
Faculty of Information Technology
Thang Long University
Hanoi, Vietnam

Vladik Kreinovich
Department of Computer Science
University of Texas at El Paso
El Paso, TX, USA

ISSN 2198-4182 ISSN 2198-4190 (electronic)
Studies in Systems, Decision and Control
ISBN 978-3-031-63928-9 ISBN 978-3-031-63929-6 (eBook)
https://doi.org/10.1007/978-3-031-63929-6

This Springer imprint is published by the registered company Springer Nature Switzerland AG
The registered company address is: Gewerbestrasse 11, 6330 Cham, Switzerland

Preface

Health issues are very important. So naturally whatever new data processing technique appears, researchers try to apply it to health issues as well. From this viewpoint, Artificial Intelligence (AI) and Computational Intelligence (CI) techniques are no exception: they have been successfully applied to medicine, and more promising applications are on the way. Applications of AI and CI techniques to health issues are the main focus of this book. The book contains applications to various health-related problems, from designing and maintaining a proper diet to enhancing hygiene to analysis of mammograms and left-right brain activity to treating diseases such as diabetes and drug addictions.

Health issues are also very delicate, because human bodies are complex organisms. No matter how interesting and promising are new ideas and new techniques, there is always a possibility of unexpected side effects. Because of this, we cannot apply untested methods to patients, we first need to test these methods on other less critical applications. Several book chapters describe such applications—whose success paves the way for these methods to be used in biomedical situations. These applications range from human/face detection to predicting student success to predicting election results to explaining the observed intensity of space light.

We hope that this book will help practitioners and researchers to learn more about computational intelligence techniques and their biomedical applications—and to further develop this important research direction.

We want to thank all the authors for their contributions and all anonymous referees for their thorough analysis and helpful comments.

The publication of this volume was partly supported by Thang Long University, Hanoi, Vietnam. Our thanks to the leadership and staff of this institution for providing crucial support. Our special thanks to Prof. Hung T. Nguyen for his valuable advice and constant support.

We would also like to thank Prof. Janusz Kacprzyk (Series Editor) and Dr. Thomas Ditzinger (Senior Editor, Engineering/Applied Sciences) for their support and cooperation with this publication.

Hanoi, Vietnam Nguyen Hoang Phuong
Hanoi, Vietnam Nguyen Thi Huyen Chau
El Paso, USA Vladik Kreinovich
December 2023

Contents

How to Estimate Unknown Unknowns: From Cosmic Light to Election Polls

Talha Azfar, Vignesh Ponraj, Vladik Kreinovich, and Nguyen Hoang Phuong

Abstract In two different areas of research—in the study of space light and in the study of voting—the observed value of the quantity of interest is twice larger than what we would expect. That the observed value is larger makes perfect sense: there are phenomena that we do not take into account in our estimations. However, the fact that the observed value is exactly twice larger deserves explanation. In this paper, we show that Laplace Indeterminacy Principle leads to such an explanation.

1 General Introduction

In two different areas of study—the study of space light and the study of elections—there is a similar puzzling phenomenon, that the observed value of the corresponding quantity is exactly twice larger that reasonable models predict. In this paper, we provide a possible common explanation for these two phenomena.

T. Azfar
Department of Electrical and Computer Engineering, University of Texas at El Paso, 500 W. University, El Paso, TX 79968, USA
e-mail: tazfar@miners.utep.edu

V. Ponraj · V. Kreinovich (✉)
Department of Computer Science, University of Texas at El Paso, 500W. University, El Paso, TX 79968, USA
e-mail: vladik@utep.edu

V. Ponraj
e-mail: vponraj@miners.utep.edu

N. H. Phuong
Artificial Intelligence Division, Information Technology Faculty, Thang Long University, Nghiem Xuan Yem Road, Hoang Mai District, Hanoi, Vietnam

N. Hoang Phuong et al. (eds.), *Machine Learning and Other Soft Computing Techniques: Biomedical and Related Applications*, Studies in Systems, Decision and Control 543, https://doi.org/10.1007/978-3-031-63929-6_1

2 First Case Study: Space Light

What is space light. Many celestial objects emit light. This is why we see many stars and galaxies, this is why other stars and galaxies can be seen via telescopes—what we see is the light they emit.

Usually, astronomers study light from visible stars and galaxies, where we can see the corresponding object. Some galaxies are too far away to be visible individually. However, since there are many of them, they contribute to the optical background that is visible by space telescopes. This background is known as *space light*.

We can estimate the expected amount of space light. We have a reasonably good understanding of:

- how galaxies are distributed in space and
- what amount of light an average galaxy emits.

Based on this information, we can estimate the amount of background light.

The observed amount of space light is twice larger than expected. Interestingly, the observed amount is almost exactly twice larger than the estimate. This means that there are some additional sources of light in the Universe; see, e.g., [1].

Natural question. That there are some unexpected sources of slight is natural. However, the fact that the observed amount of light is exactly twice larger than expected deserves explanation.

What we do in this paper. In this paper, we provide a natural explanation for this empirical fact—as well as for the similar empirical fact about elections.

3 Second Case Study: Election Polls

Election polls: reminder. To get a good understanding of how people will vote, specialists ask a random sample of people how they will vote in the forthcoming elections. This process is known as *election polls*. The percentage of people who expect to vote for a certain candidate is used as a reasonable approximation for the percentage of people who will actually vote for this candidate.

Results of election polls are approximate. Of course, percentages based on a small sample are only an approximation to the overall percentages. A natural question is: how accurate are the polls? If, based on a poll, one candidate is several points ahead, how confident are we that this candidate will win?

How is the accuracy of election polls usually estimated? It is known, from statistics (see, e.g., [3]), that:

- if we estimate the probability of an event based on the sample of size n,
- then the standard deviation σ of the corresponding accuracy is equal to $\sqrt{p \cdot (1-p)/n}$.

In particular, when we use the poll of $n = 1000$ randomly selected people to estimate the probability p of a candidate's win, then:

- for candidates with approximately equal chances, where $p \approx 0.5$,
- we get $\sigma \approx 1.7\%$.

So, with 95% confidence, this should estimate the probability with $2\sigma \approx 3.5\%$ accuracy.

Observed standard deviation is exactly twice larger. In practice, the largest deviation is twice larger then what we would expect; see, e.g., [4].

Natural question. That standard deviation is larger than expected is natural: people change their opinions, and this adds to the difference between how people answer in the poll and how they actually vote. However, the fact that the observed standard deviation is exactly twice larger than expected deserves explanation.

What we do in this paper. In this paper, we provide a natural explanation for this empirical fact—as well as for the similar empirical fact about space light.

4 Possible Explanation

Summary of what we want to explain. In both case studies, taking unknown unknowns into account doubles the corresponding value. How can we explain that?

Formulation of the general problem. In both case studies:

- we know the estimated value v, and
- we want to estimate the actual value a.

The only information that we have about a is that $a \geq v$.

Based on this information, how can we estimate a?

Let us reformulate the problem, to make it easier to answer: idea. In the above formulation, we have two real numbers: v and a. To simplify the problem, let us take into account that the numerical value of each quantity depends on the selection of a measuring unit. For example, the same height of 1.7 m takes the value 170 if we use centimeter as a measuring unit.

Let us use this idea to simplify our problem. For this purpose, let us select the unknown value a as the new measuring unit for the corresponding quantity. In terms of this new unit:

- the value a will take the form $A = 1$, and
- the value v will have the form $V = v/a$.

Thus, the above problem is reformulated as follows:

- we know that in the new unit, the actual value is 1, and
- we want to find the value V that described the estimated value in terms of this new unit.

The only information that we have about the desired value V is that $0 < V \le 1$, i.e., that the value V is that it is located on the interval [0, 1].

It is natural to use Laplace Indeterminacy Principle. We have no reason to assume that some of these values are more probable than others. So, it makes sense to assume that all these values are equally probable.

This argument is known as Laplace Indeterminacy Principle; see, e.g., [2]. Based on this argument, we conclude that the value V is uniformly distributed on the interval [0, 1].

From distribution to a single numerical estimate. We have a reasonable distribution of the set of all possible values V. What we want, however, is a single numerical estimate.

In general, if we want to represent this distribution by a single number, a reasonable choice is to select the value V_s for which the mean square deviation from the actual (unknown) value v is the smallest possible [3]. One can easily check that this V_s is the mean value of V, i.e., $V_s = 0.5$.

This conclusion indeed explains the above phenomena. We have $v/a = 1/2$. Based on this relation:

- if we know v,
- then a reasonable estimate for a is $a = 2v$.

This is exactly what we observe in the above two case studies.

Acknowledgements This work was supported in part by the National Science Foundation grants 1623190 (A Model of Change for Preparing a New Generation for Professional Practice in Computer Science), HRD-1834620 and HRD-2034030 (CAHSI Includes), EAR-2225395, and by the AT&T Fellowship in Information Technology.
It was also supported by the program of the development of the Scientific-Educational Mathematical Center of Volga Federal District No. 075-02-2020-1478, and by a grant from the Hungarian National Research, Development and Innovation Office (NRDI).

References

1. Bernal, J.L., Sato-Polito, G., Kamionkowski, M.: Cosmic optical background excess, dark matter, and line-intensity mapping. Phys. Rev. Lett. **129**, Paper 231301 (2022)
2. Jaynes, E.T., Bretthorst, G.L.: Probability Theory: The Logic of Science. Cambridge University Press, Cambridge, UK (2003)

3. Sheskin, D.J.: Handbook of Parametric and Nonparametric Statistical Procedures. Chapman and Hall/CRC, Boca Raton, Florida (2011)
4. Shirani-Mehr, H., Rotschild, D., Goel, S., Gelman, A.: Disentagling bias and variance in election polls. J. Am. Stat. Assoc. **113**(522), 607–614 (2018)

Why Bump Reward Function Works Well in Training Insulin Delivery Systems

Lehel Dénes-Fazakas, László Szilágyi, Gyorgy Eigner, Olga Kosheleva, Vladik Kreinovich, and Nguyen Hoang Phuong

Abstract Diabetes is a disease when the body can no longer properly regulate blood glucose level, which can lead to life-threatening situations. To avoid such situations and regulate blood glucose level, patients with severe form of diabetes need insulin injections. Ideally, the system should automatically decide when best to inject insulin and how much to inject. To find the optimal control, researchers applied machine learning with different reward functions. It turns out that the most effective learning occurred when they used the so-called bump function. In this paper, we provide a possible explanation for this empirical result.

L. Dénes-Fazakas
Physiological Controls Research Center and Applied Informatics and Applied Mathematics Doctoral School, Óbuda University, Bécsi Street 96/B, Budapest 1034, Hungary
e-mail: denes-fazakas.lehel@uni-obuda.hu

L. Szilágyi
Physiological Controls Research Center, Óbuda University, Bécsi Street 96/B, Budapest 1034, Hungary
e-mail: lalo@ms.sapientia.ro

Computational Intelligence Research Group, Sapientia University, Tg. Mureş, Romania

G. Eigner
Physiological Controls Research Center and Biomatics and Applied Artificial Intelligence Institute John von Neumann Faculty of Informatics, Óbuda University, Bécsi Street 96/B, Budapest 1034, Hungary
e-mail: eigner.gyorgy@uni-obuda.hu

O. Kosheleva
Department of Teacher Education, University of Texas at El Paso, 500 W. University, El Paso, TX 79968, USA
e-mail: olgak@utep.edu

V. Kreinovich (✉)
Department of Computer Science, University of Texas at El Paso, 500 W. University, El Paso, TX 79968, USA
e-mail: vladik@utep.edu

N. H. Phuong
Artificial Intelligence Division, Information Technology Faculty, Thang Long University, Nghiem Xuan Yem Road, Hoang Mai District, Hanoi, Vietnam

N. Hoang Phuong et al. (eds.), *Machine Learning and Other Soft Computing Techniques: Biomedical and Related Applications*, Studies in Systems, Decision and Control 543, https://doi.org/10.1007/978-3-031-63929-6_2

1 Formulation of the Problem

What is diabetes. All living creatures need energy to function. To many cells in a human body, energy comes from glucose that is delivered to these cells by the blood flow. The absorption of glucose into the cells is regulated by a special hormone called *insulin*.

When the body does not produce enough insulin—the illness known as *diabetes*—it hinders the ability of cells to get energy and can lead to life-threatening situations.

Need for insulin injections. To avoid dangerous situations, a natural idea is to inject insulin into the body when the insulin level becomes dangerously low—and we can detect that, since in this case the cells do not absorb the glucose and thus, the blood glucose level becomes too high.

It is desirable to have automatic insulin injections. In healthy patients, the body itself decides how much insulin is needed. In the absence of such an automatic biologic regulation, at present, the patients themselves decide when to inject insulin and how much to inject, based on some general recommendations.

The effectiveness of these general recommendations is different for different patients. It is therefore desirable to have automatic systems individually trained to becomes maximally effective for each patient. Such systems are indeed being actively developed, trained, and tested all over the world.

Empirical fact: bump reward function works the best. The purpose of the system is to keep the patient's blood sugar level x within the desired interval $[\underline{x}, \overline{x}]$. For training the automatic insulin delivery system, we can, in principle, use different reward functions. In [2], researchers compared the results of using different reward functions, and found out that the most effective is the so-called *bump* reward function that is equal to 0 outside the desired interval and to

$$b(x) = \exp\left(-\frac{c}{(x - \underline{x}) \cdot (\overline{x} - x)}\right)$$

for values x within this interval.

Natural question. A natural question is: why the bump functions works best?

What we do in this paper. In this paper, we provide a possible explanation for the effectiveness of the bump function.

2 Analysis of the Problem and the Resulting Explanation

What is a natural reward function? We want the patient to feel healthy. At each moment of time, the only information that we have about the patient is the patient's blood glucose level x. Based on this level, we can only determine the probability

$p(x)$ that the patient feels healthy. This probability is what we want to maximize, i.e., that should be our reward function.

Clearly, when the value x is outside the given interval, something is wrong, so the corresponding probability is 0 (or close to 0). So, to find an appropriate reward function, we need to find the probabilities $p(x)$ corresponding to values x from the given interval.

We need to select probabilities based on partial information. In many practical situations, probabilities are determined experimentally, as corresponding frequencies; see, e.g., [5]. However, in our case, we do not have enough statistics, so we need to select the probability distribution based on whatever information we have. For this purpose, let us recall how, in general, a probability distribution is determined based on partial information.

How probability distribution is determined based on partial information: reminder. In many practical situations, we only have partial information about probabilities.

For example, we may know that there are two possible situations, but we have no information which of the two situations is more probable. In such situations, a reasonable idea is to assign equal probability to both situations. Similarly, if we have n possible situations, and we have no reason to believe that one of them is more probable, a reasonable idea is to assign, to all of them, equal probability $1/n$. This natural idea is known as *Laplace Indeterminacy Principle*.

This principle can be described in a slightly different way. If we have two alternatives, we have an uncertainty, in which to determine which is a correct one, we need to ask one binary (= "yes"–"no") question. If we have 2^n alternatives, then we need n binary questions to uniquely determine the alternative.

- When we select equal probabilities, the average number of questions needed to determine the situation remains the same.
- However, if we selected unequal probabilities, then, on average, the number of questions becomes smaller.

For example, if we assign probability 1 to one of the alternatives and 0 to all others, we need 0 questions to find the alternative—it is the one whose probability is 1. So, in this case we kind of cheat, we insert artificial certainty where there was none.

Similarly, if we have partial information about probabilities, i.e., if there is a whole set of probability distributions that is consistent with available information, then a reasonable idea is not to cheat, not to add artificial certainty, but to preserve the original uncertainty.

- In the discrete case, a natural measure of uncertainty is the average number of binary questions that is needed to uniquely determine the alternative.
- In the continuous case, a similar natural measure is the average number of binary question that is needed to determine the unknown value x with a given accuracy $\varepsilon > 0$.

In both cases, there are distributions for which this average number of questions is smaller—but selecting them would be artificially adding certainty. What we need is the distribution that best reflects the original uncertainty, i.e., for which the average number of questions is as large as possible. It turns out that the average of question is described by Shannon's entropy

$$S = -\int f(x) \cdot \ln(f(x))\, dx, \tag{1}$$

where $f(x)$ is the corresponding probability density function; see, e.g., [1, 4]. Thus, a reasonable idea is to select, from each class of probability distributions, the distribution with the largest possible entropy. This *Maximum Entropy* approach has indeed led to many successful applications; see, e.g., [3].

Let us apply this general idea to our case: first idea. We have a class of probability distributions located on the interval $[\underline{x}, \overline{x}]$. If we do not make any assumptions about the distribution, then the only constraint on the probability density function is that the overall probability is 1:

$$\int f(x)\, dx = 1. \tag{2}$$

To maximize entropy under this constraint, it is natural to use Lagrange multiplier method, i.e., to reduce the corresponding constraint optimization problem to an equivalent unconstrained optimization problem of maximizing the function

$$-\int f(x) \cdot \ln(f(x))\, dx + \lambda \cdot \left(\int f(x)\, dx - 1 \right), \tag{3}$$

for an appropriate value λ. This value—known as *Lagrange multiplier*—is determine by the condition that the optimal function $f(x)$ satisfies the constraint (1).

The solution to this unconstrained optimization problem can be obtained by using the known fact from calculus—that the maximum of an expression is attained when all its derivatives are equal to 0. Differentiating the expression (3) with respect to each unknown $f(x)$ and equating the derivative to 0, we conclude that

$$-\ln(f(x)) - 1 + \lambda = 0,$$

i.e., that $f(x) = \exp(\lambda - 1) = \text{const}$. So, we get a uniform distribution on the desired interval – in perfect accordance with the above argument (that used Laplace Indeterminacy Principle).

Limitations of the first idea. From the mathematical viewpoint, this is reasonable, but from the viewpoint of our problem, it is not reasonable at all:

- For the uniform distribution, the probability of being healthy is exactly the same whether we are in the middle of the desired interval or close to one of its endpoints.

- However, in practice, if the value of the blood glucose level start getting closer to the threshold, this should be a sign to be alarmed—so the probability of being healthy should be smaller close to the endpoints.

This means that we cannot get a reasonable distribution if we do not impose any constraints. We need to impose some constraints if we want a reasonable result.

Second idea. We need to add constraints, and constraints reflect partial information that we have. What do we know about the probability distribution? We rarely know individual characteristic, but often, from observations, we know averages.

So, a seemingly natural idea is to add a constraint that we know the average value $\widetilde{x}$ of the quantity x:

$$\int x \cdot f(x)\,dx = \widetilde{x}. \tag{4}$$

By applying the same Lagrange multiplier method to the problem of maximizing entropy (1) under constraints (2) and (3), we arrive at the problem of optimizing the following expression:

$$-\int f(x) \cdot \ln(f(x))\,dx + \lambda_1 \cdot \left(\int f(x)\,dx - 1\right) + \lambda_2 \cdot \left(\int x \cdot f(x)\,dx - \widetilde{x}\right),$$

for which equating derivatives to 0 leads to

$$-\ln(f(x)) = 1 + \lambda_1 + \lambda_2 \cdot x = 0,$$

i.e., to $f(x) = \exp((\lambda_1 - 1) + \lambda_2 \cdot x)$.

Limitations of the second idea. The resulting formula has the same limitations as the first idea:

- we want the probability of healthiness to tend to 0 as approach the endpoints,
- but this is not happening here.

How can we modify the second idea? Let us take into account that the same physical quantity can be described by different numerical values. First, we can select a different measuring unit: e.g., the height of 2 m becomes 200 if we use centimeters. Second, we can select a different starting point: the current year 2023 can become year 2014 in Ethiopian calendar that uses a different starting date.

Finally, many quantities are ratios—e.g., blood glucose level is the ratio of the amount of glucose in blood to the corresponding amount of blood. In such cases, we can reverse the ratio and also get a meaningful description of the same quantity; for example:

- we can have velocity $v = d/t$—which is the ratio of distance to time—and we can have *slowness* $1/v = t/d$;
- we can have resistance $R = V/I$—which is the ratio of voltage to current—and we can have *conductivity* $1/R = I.V$, etc.

First try. If we simply change the measuring unit or the starting point, the situation does not change: fixing the mean value of the re-scaled quantity $k \cdot x$ or $x + x_0$ is equivalent to fixing the mean value of the quantity x itself.

What if we reverse the formula for the blood sugar level and consider the mean value of this reverse

$$\int \frac{1}{x} \cdot f(x)\, dx = \widetilde{r}$$

then the corresponding constraint optimization leads to

$$-\ln(f(x)) - 1 + \lambda_1 + \lambda_2 \cdot \frac{1}{x} = 0,$$

i.e., to

$$f(x) = \exp\left((\lambda_1 - 1) + \frac{\lambda_2}{x}\right).$$

This is still not exactly we want.

Second try leads to the desired explanation. What if we take into account both the possibility of taking a reverse *and* the probability of changing the starting points. It is reasionable to use both endpoints $\underline{x}$ and $\overline{x}$ as starting points. Thus:

- we get the re-scaled values $x - \underline{x}$ and $x - \overline{x}$ (or, better, $\overline{x} - x$, to keep the values non-negative), and
- reversing these re-scaled values leads to the following two constraints:

$$\int \frac{1}{x - \underline{x}} \cdot f(x)\, dx = \widetilde{r}_- \tag{5}$$

and

$$\int \frac{1}{\overline{x} - x} \cdot f(x)\, dx = \widetilde{r}_+. \tag{6}$$

Maximizing the expression (1) under constraints (5) and (6) leads to the following unconstrained optimization problem:

$$-\int f(x) \cdot \ln(f(x))\, dx + \lambda_1 \cdot \left(\int f(x) - 1\right) +$$

$$\lambda_- \cdot \left(\int \frac{1}{x-\underline{x}} \cdot f(x)\,dx - \widetilde{r}_-\right) + \lambda_+ \cdot \left(\int \frac{1}{\overline{x}-x} \cdot f(x)\,dx - \widetilde{r}_+\right).$$

For this expression, equating its derivatives to 0 leads to

$$\ln(f(x)) = -1 + \lambda_1 + \frac{\lambda_-}{x-\underline{x}} + \frac{\lambda_+}{\overline{x}-x}. \quad (7)$$

Here:

- the value λ_- reflects the importance of the left endpoint of the desired interval,
- while the value λ_+ reflects the uncertainty of the right endpoint.

Both endpoints are important. Going beyond each of these two endpoints can be life-threatening, and we have no reason to assume that one of the endpoints is more important. Thus, in line with the Laplace Indeterminacy Principle, it makes sense to assume that these two values are equal: $\lambda_- = \lambda_+$. In this case, the formula (7) takes the form

$$\ln(f(x)) = \text{const} + \frac{\text{const}}{(x-\underline{x}) \cdot (\overline{x}-x)}.$$

So, we get exactly the bump function expression for the probability values $f(x)$!

Conclusion. Thus, we have indeed explained the effectiveness of the bump reward function: the explanation is that this function naturally follows from first principles.

Acknowledgements This work was supported in part by the National Science Foundation grants 1623190 (A Model of Change for Preparing a New Generation for Professional Practice in Computer Science), HRD-1834620 and HRD-2034030 (CAHSI Includes), EAR-2225395, and by the AT&T Fellowship in Information Technology.
It was also supported by the program of the development of the Scientific-Educational Mathematical Center of Volga Federal District No. 075-02-2020-1478, and by a grant from the Hungarian National Research, Development and Innovation Office (NRDI).

References

1. Chokr, B., Kreinovich, V.: How far are we from the complete knowledge: complexity of knowledge acquisition in Dempster-Shafer approach. In: Yager, R.R., Kacprzyk, J., Pedrizzi, M. (eds.) Advances in the Dempster-Shafer Theory of Evidence, pp. 555–576. Wiley, N.Y. (1994)
2. Dénes-Fazakas, L., Siket, M., Szilágyi, L., Eigner, G., Kovacs, L.: Investigation of reward functions for controlling blood glucose level using reinforcement learning. Proceedings of the IEEE 17th International Symposium on Applied Computational Intelligence and Informatics SACI 2023, pp. 387–392. Timişoara, Romania (2023)
3. Jaynes, E.T., Bretthorst, G.L.: Probability Theory: The Logic of Science. Cambridge University Press, Cambridge, UK (2003)
4. Nguyen, H.T., Kreinovich, V., Wu, B., Xiang, G.: Computing Statistics under Interval and Fuzzy Uncertainty. Springer, Berlin, Heidelberg (2012)
5. Sheskin, D.J.: Handbook of Parametric and Nonparametric Statistical Procedures. Chapman and Hall/CRC, Boca Raton, Florida (2011)

We Can Always Reduce a Non-linear Dynamical System to Linear—At Least Locally—But Does It Help?

Orsolya Csiszár, Gábor Csiszár, Olga Kosheleva, Vladik Kreinovich, and Nguyen Hoang Phuong

Abstract Many real-life phenomena are described by dynamical systems. Sometimes, these dynamical systems are linear. For such systems, solutions are well known. In some cases, it is possible to transform a nonlinear system into a linear one by appropriately transforming its variables, and this helps to solve the original nonlinear system. For other nonlinear systems—even for the simplest ones—such transformation is not known. A natural question is: which nonlinear systems allow such transformations? In this paper, we show that we can always reduce a nonlinear system to a linear one—but, in general, it does not help, since the complexity of finding such a reduction is exactly the same as the complexity of solving the original nonlinear system.

O. Csiszár
Research Center for Complex Systems, Aalen University, Beethovenstrasse 1, 73431 Aalen, Germany
e-mail: orsolya.csiszar@hs-aalen.de; csiszar.orsolya@nik.uni-obuda.hu

Institute of Applied Mathematics John von Neumann Faculty of Informatics, Óbuda University, Bécsi Street 96/B, Budapest 1034, Hungary

G. Csiszár
Biomatics and Applied AI Institute John von Neumann Faculty of Informatics, Óbuda University, Bécsi Street 96/B, Budapest 1034, Hungary
e-mail: csiszar.gabor@uni-obuda.hu

O. Kosheleva
Department of Teacher Education, University of Texas at El Paso, 500 W. University, El Paso, TX 79968, USA
e-mail: olgak@utep.edu

V. Kreinovich (✉)
Department of Computer Science, University of Texas at El Paso, 500 W. University, El Paso, TX 79968, USA
e-mail: vladik@utep.edu

N. Hoang Phuong
Artificial Intelligence Division, Information Technology Faculty, Thang Long University, Nghiem Xuan Yem Road, Hoang Mai District, Hanoi, Vietnam

N. Hoang Phuong et al. (eds.), *Machine Learning and Other Soft Computing Techniques: Biomedical and Related Applications*, Studies in Systems, Decision and Control 543, https://doi.org/10.1007/978-3-031-63929-6_3

1 Formulation of the Problem

Dynamical systems are ubiquitous. One of the main objectives of science is to predict the future state of different systems. The state of a system can be described by the values $x_1, \ldots, x_n$ of different quantities that characterize this system. In many cases, the current state of the system uniquely determines the rate $\dot{x}_i$ with which each of the values changes:

$$\dot{x}_i = f_i(x_1, \ldots, x_n), \quad 1 \le i \le n. \tag{1}$$

Such a system of differential equations is known as a *dynamical system.*

For a dynamical system, the current state of the system uniquely determines its state at any future (or past) moment of time; see, e.g., [1, 2].

In many cases, we have linear dynamical systems. In many practical situations, the changes are relatively small. In such cases, for all moments of time, the values x_i are close to their initial values $x_i^{(0)}$. In such cases, it is convenient to describe the state of the system by the differences $y_i \stackrel{\text{def}}{=} x_i - x_i^{(0)}$ for which $x_i = x_i^{(0)} + y_i$ and, thus, $\dot{x}_i = \dot{y}_i$. In terms of these differences, the Eq. (1) take the form

$$\dot{y}_i = f\left(x_1^{(0)} + y_1, \ldots, x_n^{(0)} + y_n\right). \tag{2}$$

Since the values y_i are relatively small, we can safely ignore terms that are quadratic or of higher order in terms of y_i. For example, if $y_i \approx 10\%$, then $y_i^2 \approx 1\%$, which is much smaller than y_i. Thus, we can expand the right-hand side of the formula (2) in Taylor series in terms of y_i and keep only linear terms in this expansion. In this case, the right-hand side of the system (2) becomes linear in $y_1, \ldots, y_n$;

$$\dot{y}_i = a_i + \sum_{j=1}^{n} a_{i,j} \cdot y_j, \tag{3}$$

for some coefficients a_i and $a_{i,j}$. Such dynamical systems are called *linear.*

Linear systems are easy to solve. There are known formulas for solving linear systems.

Namely, in the absence of the constant terms a_i, the general solution to this system is a linear combination of the terms $t^k \cdot \exp(\lambda \cdot t)$, where:

- λ is an eigenvalue of the matrix $a_{i,j}$, and
- k is a non-negative integer which is smaller than the multiplicity of this eigenvalue.

A simple modification of this formula—in most cases, by adding constants—takes care of the general case of (3).

Sometimes, we can reduce a non-linear system to a linear one. In some cases, the differences y_i are not small, so we have to consider the original nonlinear dynamical system (1). For such systems, there is no general way to solve it other than to solve it numerically. However, in some cases, it is possible to reduce a nonlinear system to a linear one and thus, to get an explicit solution.

A known example—for which such a reduction is possible—is the following system:

$$\dot{x}_1 = a \cdot x_1, \tag{4}$$

$$\dot{x}_2 = b \cdot (x_2 - x_1^2). \tag{5}$$

In this system, the first equation (4) is linear, but the second (5) is nonlinear.

This system can be reduced to a system of linear equations if we add the third variable $x_3 = x_1^2$. In this case, the second equation (5) becomes linear:

$$\dot{x}_2 = b \cdot x_2 - b \cdot x_3, \tag{6}$$

and for the derivative of x_3, we have

$$\dot{x}_3 = \frac{d}{dt}(x_1^2) = 2 \cdot x_1 \cdot \dot{x}_1.$$

Substituting the expression (4) for $\dot{x}_1$ into this formula, we get

$$\dot{x}_3 = 2 \cdot x_1 \cdot (b \cdot x_1) = 2b \cdot x_1^2.$$

By definition of x_3 as x_1^2, this implies that for the rate of change of the third variables, we also have a linear equation

$$\dot{x}_3 = 2b \cdot x_3. \tag{7}$$

For other nonlinear equations, such a reduction is not known. Let us consider the simplest possible nonlinear dynamical system.

- The more variables, the more complex the system, so the simplest system should have a single variable x_1—and thus, the dynamical system consists of a single differential equation $\dot{x}_1 = f_1(x_1)$. To find the simplest of such systems, we need to select the simplest possible nonlinear function $f_1(x_1)$.
- The simplest possible functions are linear, and the simplest nonlinear functions are quadratic. The simplest-to-compute quadratic function is simply the expression x_1^2 that can be computed by a single multiplication. So, the simplest possible nonlinear system is the equation

$$\dot{x}_1 = x_1^2. \tag{8}$$

For this system—as well as for other differential equations with quadratic right-hand sides—no reduction to linear systems is known.

Natural questions. For some nonlinear systems, there is a known reduction to linear ones, and this reduction helps solve the system. This fact leads to the following two natural questions:

- which nonlinear systems can be reduced to linear ones? and
- if such a reduction is possible, can it help solve the original nonlinear system?

What we do in this paper. In this paper, we provide answers to both questions. Namely, we show:

- that *every* nonlinear system can be reduced to a linear one—at least locally, but
- that this does not help us solve the original nonlinear system—finding this reduction is as complicated as actually solving the system.

2 Our Answers

Notation. As we have mentioned, in general:

- once we know the Eq. (1) and we know the state $x = (x_1, \ldots, x_n)$ of the system at some moment t_0,
- we can uniquely determine its state at any other moment of time (at least locally, i.e., for times t sufficiently close to t_0).

Since the system (1) does not explicitly include time, the transition between the state at moment t_0 and the state at moment t depends only on the difference $t - t_0$. Let us denote the state at moment t corresponding to the state x at moment t_0 by $T_{t-t_0}(x)$. The components of the state $T_{t-t_0}(x)$ will be denoted by $T_{t-t_0,1}(x), \ldots, T_{t-t_0,n}(x)$. In these terms, the state $x = (x_1, \ldots, x_n)$ is transformed into the state

$$(z_1, \ldots, z_n) = (T_{t-t_0,1}(x_1, \ldots, x_n), \ldots, T_{t-t_0,n}(x_1, \ldots, x_n)).$$

Clearly, if we let the system evolve for time t and then again for time t', this is equivalent to letting it evolve for time $t + t'$, i.e., we have

$$T_{t+t'}(x) = T_{t'}(T_t(x)). \tag{9}$$

How to reduce a nonlinear system to linear: idea. Let us consider the cases when the values of the last variable x_n are close to some number $x_n^{(0)}$. In general, unless the system is degenerate and the value x_n does not change at all, the value x_n changes with time. For each tuple $(x_1, \ldots, x_{n-1}, x_n)$, we can look for the time τ at which the value x_n is equal to $x_n^{(0)}$, i.e., at which

$$T_{\tau,n}(x_1, \ldots, x_n) = x_n^{(0)}. \tag{10}$$

To find the unknown time τ, we have one Eq. (9) with one unknown τ. In general, if the number of equations is equal to the number of unknowns, the system has a unique solution—at least locally. Thus, we can always find such time τ.

Towards the resulting reduction. Now, instead of the original variables $x_1, \ldots, x_n$, we can use the new variables $(y_1, \ldots, y_n)$ in which:

- the value y_n is equal to the time τ determined by the formula (9), and
- the values $y_1, \ldots, y_i, \ldots, y_{n-1}$ are equal to

$$y_i = T_{\tau,i}(x_1, \ldots, x_n). \tag{11}$$

How will the variables y_i change with time? Due to formulas (9) and (10), we have

$$\left(y_1, \ldots, y_{n-1}, x_n^{(0)}\right) = T_\tau(x_1, \ldots, x_n),$$

i.e., since $\tau = y_n$, we have

$$\left(y_1, \ldots, y_{n-1}, x_n^{(0)}\right) = T_{y_n}(x_1, \ldots, x_n). \tag{12}$$

Thus, to describe x in terms of y, it is sufficient to trace the changes of the state $\left(y_1, \ldots, y_{n-1}, x_n^{(0)}\right)$ back in time for the period $\tau = y_n$:

$$(x_1, \ldots, x_n) = T_{-y_n}\left(y_1, \ldots, y_{n-1}, x_n^{(0)}\right). \tag{13}$$

In time t, the state x turns into $T_t(x)$. By applying the transformation T_t to both sides of the formula (11) and taking into account the formula (8), we get

$$T_t(x_1, \ldots, x_n) = T_{-(y_n - t)}\left(y_1, \ldots, y_{n-1}, x_n^{(0)}\right). \tag{14}$$

Thus, by definition of the y-state, the y-state corresponding to $T_t(x)$ has the form $(y_1, \ldots, y_{n-1}, y_n - t)$. Thus, with time:

- the new variables $y_1, \ldots, y_{n-1}$ do not change at all, and
- the variable y_n linearly decreases with time.

So, we indeed have the desired reduction. Let us summarize it.

Resulting reduction. For each state $x = (x_1, \ldots, x_n)$, let us define the new variables $y = (y_1, \ldots, y_n)$ as follows: y_n is the solution to the equation

$$T_{y_n,n}(x_1, \ldots, x_n) = x_n^{(0)}, \tag{15}$$

and for all $i < n$, we have

$$y_i = T_{y_n,i}(x_1, \dots, x_n). \tag{16}$$

For the new variables, the dynamical system has the following form:

$$\dot{y}_1 = \dots = \dot{y}_{n-1} = 0, \quad \dot{y}_n = -1. \tag{17}$$

From the new variables y_i, we can get back to the original variables x_i by using the formula (12).

Example: derivation. Let us illustrate our reduction on the example of the simplest nonlinear dynamical systems $\dot{x}_1 = x_1^2$. This system is easy to solve. Indeed, we start with the original equation:

$$\frac{dx_1}{dt} = x_1^2.$$

We then separate the variables by multiplying both sides by dt and dividing both sides by x_1^2. This results in:

$$\frac{dx_1}{x_1^2} = dt.$$

If we integrate both sides of this equation, we get

$$-\frac{1}{x_1} + C = t,$$

where C denotes the integration constant. Thus,

$$\frac{1}{x_1} = C - t,$$

and so,

$$x_1(t) = \frac{1}{C - t}. \tag{18}$$

Let us describe the corresponding function $T_t(x_1)$ that describes the transition from the state at moment 0 to the state at moment t. At moment 0, the formula (17) leads to

$$x_1(0) = \frac{1}{C},$$

so

$$C = \frac{1}{x_1(0)}.$$

Substituting this value C into the formula (17), we conclude that

$$x_1(t) = \frac{1}{\dfrac{1}{x_1(0)} - t} = \frac{x_1(0)}{1 - x_1(0) \cdot t}.$$

Thus,

$$T_t(x_1) = \frac{x_1}{1 - x_1 \cdot t}. \tag{19}$$

Let us take $x_1^{(0)} = 1$. Then, the value y_1 should be determined by the Eq. (14) which, in this case, has the form

$$\frac{x_1}{1 - x_1 \cdot y_1} = 1,$$

hence $x_1 = 1 - x_1 \cdot y_1$, so $x_1 \cdot y_1 = 1 - x_1$, and

$$y_1 = \frac{1 - x_1}{x_1} = \frac{1}{x_1} - 1. \tag{20}$$

For the new variable y_1, the dynamical system takes a linear form: $\dot{y}_1 = -1$.

Once we know new state y_1, we can reconstruct the original state x_1 by using the formula (12), which in this case takes the form

$$x_1 = T_{-y_1}(1) = \frac{1}{1 - 1 \cdot (-y_1)} = \frac{1}{1 + y_1}. \tag{21}$$

Example: summary. For the nonlinear system $\dot{x}_1 = x_1^2$, we can take the new variable

$$y_1 = \frac{1}{x_1} - 1$$

for which

$$x_1 = \frac{1}{1 + y_1}.$$

For this new variable, the dynamical system has the form $\dot{y} = -1$.

Does this reduction help solve the original nonlinear system? Not really, since the reduction uses the solution $T_t(x)$ of the system.

But can we have a reduction that does help solve the system? Not really:

- if we know the solution, then, as we have shown, we can easily find the appropriate reduction;
- vice versa, if we know a reduction to a linear system, then, since solutions to linear systems are known, we can easily find the solution to the original nonlinear system.

Because of this, the complexity of finding a reduction to a linear system is exactly the same as the complexity of solving the original nonlinear system.

Acknowledgements This work was supported in part by the National Science Foundation grants 1623190 (A Model of Change for Preparing a New Generation for Professional Practice in Computer Science), HRD-1834620 and HRD-2034030 (CAHSI Includes), EAR-2225395, and by the AT&T Fellowship in Information Technology.
It was also supported by the program of the development of the Scientific-Educational Mathematical Center of Volga Federal District No. 075-02-2020-1478, and by a grant from the Hungarian National Research, Development and Innovation Office (NRDI).

References

1. Feynman, R., Leighton, R., Sands, M.: The Feynman Lectures on Physics. Addison Wesley, Boston, Massachusetts (2005)
2. Thorne, K.S., Blandford, R.D.: Modern Classical Physics: Optics, Fluids, Plasmas, Elasticity, Relativity, and Statistical Physics. Princeton University Press, Princeton, New Jersey (2021)

How to Best Retrain a Neural Network if We Added One More Input Variable

Saeid Tizpaz-Niari and Vladik Kreinovich

Abstract Often, once we have trained a neural network to estimate the value of a quantity y based on the available values of inputs $x_1, \ldots, x_n$, we learn to measure the values of an additional quantity that have some influence on y. In such situations, it is desirable to re-train the neural network, so that it will be able to take this extra value into account. A straightforward idea is to add a new input to the first layer and to update all the weights based on the patterns that include the values of the new input. The problem with this straightforward idea is that while the result is a minor improvement, such re-training will take a lot of time, almost as much as the original training. In this paper, we show, both theoretically and experimentally, that in such situations, we can speed up re-training—practically without decreasing resulting accuracy—if we only update some weights.

1 Formulation of the Problem

Need for machine learning: a brief reminder. In many real-life situations, the value of a quantity y is, to a large extent, determined by the values of related quantities $x_1, \ldots, x_n$. This situation is typical for predictions, where the future value y of some quantity (e.g., tomorrow's temperature) is largely determined by today's values of temperature, wind speed, humidity, etc., at this location and at nearby locations.

In some cases, we know explicit formulas—or at least effective algorithms—for determining y based on $x_1, \ldots, x_n$; this is, e.g., the case of celestial mechanics, where we can predict solar eclipses hundreds of years ahead. However, in many other cases, no such formula or algorithm is known. In such cases, all we have is many (P) cases when we know both the values $x_1^{(p)}, \ldots, x_n^{(p)}$ of the input and the value $y^{(p)}$ of the

S. Tizpaz-Niari · V. Kreinovich (✉)
Department of Computer Science, University of Texas at El Paso, 500 W. University El Paso, El Paso, TX 79968, USA
e-mail: vladik@utep.edu

S. Tizpaz-Niari
e-mail: saeid@utep.edu

N. Hoang Phuong et al. (eds.), *Machine Learning and Other Soft Computing Techniques: Biomedical and Related Applications*, Studies in Systems, Decision and Control 543, https://doi.org/10.1007/978-3-031-63929-6_4

output. Based on these patterns $\left(x_1^{(p)}, \ldots, x_n^{(p)}, y^{(p)}\right)$, we need to find an algorithm $f(x_1, \ldots, x_n)$ for which, for each of these patterns $p = 1, \ldots, P$, we have

$$y^{(p)} \approx f\left(x_1^{(p)}, \ldots, x_n^{(p)}\right).$$

Finding such algorithm based on the given patterns is known as *regression* in statistics and as *machine learning* in computer science.

Deep learning: a brief reminder. At present, one of the most effective machine learning tools is *deep learning* that uses deep (multi-layer) networks of (artificial) neurons. In general, in a neural network, each neuron—except for the very last one—transforms its inputs $z_1, \ldots, z_m$ into the output

$$t = s(w_1 \cdot z_1 + \cdots + w_m \cdot z_m + w_0),$$

where:

- $s(z)$ is a given nonlinear function called *activation function*, and
- w_i are numerical parameters called *weights*.

The last neuron simply returns the value $y = w_1 \cdot z_1 + \cdots + w_m \cdot z_m + w_0$, without any additional nonlinear transformation.

Neurons are usually divided into layers:

- Neurons from the first layer process the original data $x_1, \ldots, x_n$ (e.g., the original measurement results).
- Neurons from the second layer use, as inputs, outputs of the neurons from the first layer.
- In general, neurons from the $(\ell + 1)$-st layer use, as inputs, outputs of the neurons from the ℓth layer.

The output(s) of the neuron(s) of the last layer is the final answer that is returned to the user.

- In some cases, there are only two layers; such neural networks are called *shallow*. This was a traditional way neural networks were used in the past, see, e.g., [3]. Shallow neural networks usually use the activation function $s(z) = 1/(1 + \exp(-z))$ called *sigmoid*.
- When a neural network contains a reasonably large number of layers, it is called *deep*. In deep learning, the neurons use the activation function $s(z) = \max(0, z)$. This function is known as *rectified linear unit*, or *ReLU*, for short.

The weights w_i are selected so as to minimize, for each of the given patterns p, some measure of difference between the desired output $y^{(p)}$ and the result of applying the network with current weights to this pattern's inputs $x_1^{(p)}, \ldots, x_n^{(p)}$. This minimization is usually performed by gradient descent, with a special algorithm called *backpropagation* that speed up the computation of the corresponding gradients.

Terminological comment. The only layer whose results are visible to the user is the last layer. Because of this, all other layers are known as *hidden* layers. In these terms:

- a shallow neural network contains only one hidden layer, while
- a deep neural network contains a reasonable large number of hidden layers.

Computational comment. Usually, as we have mentioned, the inputs to each neuron on the $(\ell + 1)$st layer come from outputs of the neurons on the previous level ℓ. However, sometimes, it is convenient to add, as additional inputs for this neuron, some outputs from the previous layers $\ell - 1, \ell - 2$, all the way to (some of the) input signals $x_1, \ldots, x_n$. Such neural networks are called *residual*. A simple explanation of why residual networks are sometimes more effective is given, e.g., in [6].

Often, we need to add an extra variable. The list of inputs $x_1, \ldots, x_n$ is usually limited to quantities whose values are available—and about which we know that they affect the value of the quantity y. This list may miss some quantities whose use may lead to a better prediction of y:

- either because we do not know that the use of this extra variable will be useful,
- or because we do not know how to measure the corresponding quantity.

Later on, we may learn that this new variable is useful—e.g., we may learn that one of the obscure numbers included in a routine blood test may help to better diagnose some disease.

In such situation, we face the following problem:

- we already have a neural network trained to predict the desired value y based on the values of the quantities $x_1, \ldots, x_n$;
- we also have some patterns $\left(x_1^{(p)}, \ldots, x_n^{(p)}, x_{n+1}^{(p)}, y^{(p)}\right)$ that include the values of the new variable x_{n+1};
- we would like to have a neural network trained to predict y based on the values of all available quantities $x_1, \ldots, x_n, x_{n+1}$, including the new quantity x_{n+1}.

A straightforward idea, its limitations, and the resulting problem. A natural idea is to start with the current trained neural network (with n inputs):

- we add a new input to the first layer, with (e.g.) random weights from x_{n+1} to all the neurons in the first layer, and then
- we use backpropagation to train the resulting network based on the newly available patterns.

The problem is that, in general, training takes a long time. There is not much that we can do in general, when we start "from scratch". However, in our situation, we are not starting from scratch, we start with the model that already has good predictions, we are just making small improvements to these predictions.

In general, in numerical computations, the knowledge of an approximate solution enables us to speed up computations in comparison with situations when no approximate solution is known and we need to start from scratch. A natural question is: can we speed up this re-training?

What we do in this paper. In this paper, we show that it is indeed possible to speed up the re-training caused by adding an extra quantity.

2 Analysis of the Problem

Main idea. The only way to speed up re-training is *not* to perform the *full* backpropagation, i.e., in effect, *not* to reach the absolute minimum of the objective function. In other words, if we do not use full re-training, the resulting network may be not as accurate as it could be.

This may not be so bad if we take into account that neural networks provide, in general, only an approximate description of the actual dependence. So:

- if the inaccuracy caused by not doing the full re-training is comparable with the usual approximation errors of the usual neural networks,
- then this minor inaccuracy is quite acceptable.

This acceptability can be explained on a simple example:

- if we measure the car's weight with an accuracy of 1 kg, and
- then we add load to it to find the total weight of the loaded car,

it does not make sense to measure the weight of the load with a 1 g accuracy: when we weigh the load, the accuracy of 1 kg is quite sufficient.

To use this idea, let us analyze how accurately can a generic function be approximated by a neural network. For this purpose, let us first analyze how accurately a function can be approximated in general.

How accurately can a function be approximated in general? In real life, most dependencies are analytical—or at least well described by analytical functions; see, e.g., [4, 7]. A natural way to approximate such functions is to expand the corresponding expression into Taylor series and to keep the first few terms in this expansion. If we keep only linear terms, we get an expression with $n + 1$ parameters a_i:

$$f(x_1, \ldots, x_n) = a_0 + \sum_{1 \le i \le n} a_i \cdot x_i.$$

If we also retain quadratic terms, then we get an expression with $O(n^2)$ parameters:

$$f(x_1, \ldots, x_n) = a_0 + \sum_{1 \le i \le n} a_i \cdot x_i + \sum_{1 \le i \le j \le n} a_{i,j} \cdot x_i \cdot x_j,$$

If we also retain cubic terms, we need $O(n^3)$ parameters:

$$f(x_1, \ldots, x_n) = a_0 + \sum_{1 \le i \le n} a_i \cdot x_i + \sum_{1 \le i \le j \le n} a_{i,j} \cdot x_i \cdot x_j +$$

$$\sum_{1\le i\le j\le k\le n} a_{i,j,k}\cdot x_i\cdot x_j\cdot x_k. \tag{1}$$

To get a more accurate representation, we can also retain 4th order terms, this will require $O(n^4)$ parameters:

$$f(x_1,\ldots,x_n)=a_0+\sum_{1\le i\le n} a_i\cdot x_i+\sum_{1\le i\le j\le n} a_{i,j}\cdot x_i\cdot x_j+$$

$$\sum_{1\le i\le j\le k\le n} a_{i,j,k}\cdot x_i\cdot x_j\cdot x_k+\sum_{1\le i\le j\le k\le \ell\le n} a_{i,j,k,\ell}\cdot x_i\cdot x_j\cdot x_k\cdot x_\ell.$$

How much accuracy can we attain with different approximation schemes? We would like to know how accurate is the approximation provided by a deep neural network. To answer this question, let us analyze the general situation: what accuracy can we attain if we use a general approximation scheme

$$F(x_1,\ldots,x_n,c_1,\ldots,c_N)$$

with N parameters?

Clearly, potential accuracy depends on the number of parameters: the larger the number of parameters, the more accurately we can approximate different functions. To analyze how exactly the approximation accuracy depends on the number of parameters, let us take into account that we described approximation accuracy in terms of the numbers of terms in the Taylor expansion that are accurately reproduced. From this viewpoint, a natural idea is to also expand the general approximating expression in Taylor series:

$$F(x_1,\ldots,x_n,c_1,\ldots,c_N)=A_0(c_1,\ldots,c_N)+\sum_{1\le i\le n} A_i(c_1,\ldots,c_N)\cdot x_i+\cdots$$

If we have $N=n+1$, then, in principle, we can perfectly fit all linear terms in the Taylor expansion of the function $f(x_1,\ldots,x_n)$ that we want to approximate. Indeed, for this to be possible, we need to satisfy the following $n+1$ equations:

$$A_0(c_1,\ldots,c_N)=a_0,\quad A_1(c_1,\ldots,c_N)=a_1,\quad \ldots,\quad A_n(c_1,\ldots,c_N)=a_n.$$

Here, we have $N=n+1$ equations to determine $N=n+1$ unknowns $c_1,\ldots,c_N$. In general:

- If the number of equations is equal to (or smaller than) the number of unknowns, the system *has* a solution. This is true in the generic case of a linear system, and—due to the possibility of linearization—it is usually true for nonlinear systems as well.

- On the other hand, if we have more equations than unknowns, then, in general, the corresponding system *does not have* a solution—even in the general case of linear equations.

In our case, the number of equations is equal to the number of unknowns. Thus, the above system of equations has a solution. Hence, in principle, with an approximating scheme with $N = n + 1$ parameters, we can accurately fit linear terms in the expansion of the function $f(x_1, \ldots, x_n)$.

On the other hand, with this approximation scheme, we cannot, in general, fit all quadratic terms as well. Indeed, to exactly fit all these terms, we will also need to satisfy additional equations $A_{i,j}(c_1, \ldots, c_N) = a_{i,j}$ for all i and j for which $1 \le i \le j \le n$—to the total of $c \cdot n^2$ equations. So, to find the coefficients c_i, we would need to satisfy $c \cdot n^2$ equations. But since we only have $n + 1$ unknowns, the number of equations is much larger than the number of unknowns—so this system does not have a solution.

So, if we use an approximating scheme with $N = n + 1$ parameters, we fit all linear terms, but this approximation ignores quadratic (and higher order) terms.

Similarly, if we use an approximation scheme with $N = n^2$ parameters c_i, then we can always have a solution to $O(n^2)$ equations corresponding to matching all the $O(n^2)$ coefficients a_0, a_i, and $a_{i,j}$: since in this case, the number of equations is smaller than the number of unknowns. However, in this case, we cannot exactly fit cubic terms—this would mean satisfying $c \cdot n^3$ equations, and we have much fewer unknown than that: $n^2 \ll c \cdot n^3$. So, with $N = n^2$ parameters:

- we can fit all quadratic terms, and
- the largest ignored terms are cubic terms.

Same arguments show that if we use an approximation scheme with $N = n^3$ parameters, then:

- we can perfectly fit all cubic terms, and
- the largest ignored terms are 4th order terms, etc.

From this viewpoint, to find out how accurate is the approximation provided by a deep neural network, it is necessary to analyze how many parameters this approximation scheme has.

How many parameters does the deep learning approximation has: a rough estimate. In general, in a deep neural network, to process a reasonably large number n of inputs, we use a reasonably large number of layers, with each layer containing a reasonably large number of neurons. In this phrase, we use the term "reasonably large" three times:

- to describe the number of inputs,
- to describe the number of layers, and
- to describe the number of neurons in each layer.

In general, these numbers may be different. However, for the purpose of providing a rough estimate, let us assume that these numbers are equal. In other words, we assume that we have n layers, each of which has exactly n neurons in this layer. (The only exception of the last layer: since we want to output a single number y, the last layer contains only one neuron.)

How many parameters do we have here? In a deep neural networks, parameters are weights. Each of n inputs x_i can becomes an input to each of n neurons j in the first layer, with some weight $w_{j,i}$. Thus, to fully describe all the weights of the first layer, we need to describe $n \cdot n = n^2$ parameters. Similarly, for each layer k, the output of each of n neurons i from this layer can serve as the input to each of n neurons j in the next layer, with some weights $w_{j,i}$—so, again, we have n^2 parameters. (Here too, the exception is the last layer—it only has n parameters.)

So, we have n layers, and to describe each layer, we need n^2 parameters. Thus, overall, we need $n \cdot n^2 = n^3$ parameters to describe a deep neural network.

So what is the resulting accuracy. We know that deep neural network contains about n^3 parameters. Thus, based on our general analysis of approximation schemes, we can conclude that:

- a deep neural network can perfectly describe all cubic terms in the expansion of the desired function $f(x_1, \ldots, x_n)$,
- while the 4th order terms will be ignored.

In other words, we get an approximate expression of the type (1).

Comment. If, instead of a deep neural network, we had a "shallow" network, with only one hidden layer, with n neurons in this layer, then this network would contain:

- n^2 parameters relating each of n inputs with each of n neurons in the hidden layer,
- n free terms w_0 of n neurons in the hidden layer,
- n parameters relating each neuron from the hidden later to the output neuron, and
- a free term of the output neuron,

to the total of $n^2 + 2n + 1 = (n + 1)^2$ parameters. This number is larger than the number of coefficient in the general quadratic expression, but much smaller than the number of coefficients in the general cubic expression. Thus, this shallow network would be able to fit quadratic terms—but already cubic terms will not be covered.

What if we add an extra variable x_{n+1}? If we add an extra variable x_{n+1}, then, instead of the original expression (1), we have a similar expression, but with $n + 1$ variables instead of the original n ones:

$$f(x_1, \ldots, x_n, x_{n+1}) = a_0 + \sum_{1 \le i \le n+1} a_i \cdot x_i + \sum_{1 \le i \le j \le n+1} a_{i,j} \cdot x_i \cdot x_j +$$

$$\sum_{1 \le i \le j \le k \le n+1} a_{i,j,k} \cdot x_i \cdot x_j \cdot x_k, \tag{2}$$

i.e., if we separate the dependence on x_{n+1}:

$$f(x_1, \ldots, x_n, x_{n+1}) = a_0 + \sum_{1\le i\le n} a_i \cdot x_i + a_{n+1} \cdot x_{n+1} + \sum_{1\le i\le j\le n} a_{i,j} \cdot x_i \cdot x_j +$$

$$\sum_{1\le i\le n} a_{i,n+1} \cdot x_i \cdot x_{n+1} + a_{n+1,n+1} \cdot x_{n+1}^2 + \sum_{1\le i\le j\le k\le n} a_{i,j,k} \cdot x_i \cdot x_j \cdot x_k +$$

$$\sum_{1\le i\le j\le n} a_{i,j,n+1} \cdot x_i \cdot x_j \cdot x_{n+1} +$$

$$\sum_{1\le i\le n} a_{i,n+1,n+1} \cdot x_i \cdot x_{n+1}^2 + a_{n+1,n+1,n+1} \cdot x_{n+1}^3.$$

What does the neural network learn if it can only use the first n inputs? Let us analyze what exactly the original neural network learns when it only uses the first n inputs $x_1, \ldots, x_n$ during training.

If the neural network could use all $n+1$ inputs, then, for each combination of inputs $(x_1, \ldots, x_n, x_{n+1})$, the neural network would see the corresponding value (2). Thus, if we had a sufficient number of patterns and spend sufficient time on training, we would have the network learning this expression (2).

In situations when the neural network does not have access to the value x_{n+1}, then, for each combination of inputs $(x_1, \ldots, x_n)$, the neural network will have several slightly different outputs

$$y^{(p)} = f\left(x_1, \ldots, x_n, x_{n+1}^{(p)}\right) \tag{3}$$

corresponding to different values $x_{n+1}^{(p)}$ $(1 \le p \le P_0)$ of the extra quantity x_{n+1}. The result $F(x_1, \ldots, x_n)$ of the training is determined by the condition that an appropriate quantity $Q(y^{(1)}, \ldots, y^{(P_0)}, F(x_1, \ldots, x_n))$—that describes how close is the value $F(x_1, \ldots, x_n)$ to all the observed outputs $y^{(1)}, \ldots, y^{(P_0)}$—should be minimized. For example, if we minimize the least squares difference—as was typical for shallow neural networks, i.e., minimize the expression

$$Q(y^{(1)}, \ldots, y^{(P_0)}, F(x_1, \ldots, x_n)) = \sum_{1\le p\le P_0} \left(y^{(p)} - F(x_1, \ldots, x_n)\right)^2,$$

then we get

$$F(x_1, \ldots, x_n) = \frac{1}{P_0} \cdot \sum_{1\le p\le P_0} y^{(p)}.$$

In general, if we use a generic minimized expression Q, then we get a more complex expression of $F(x_1, \ldots, x_n)$ as a function of the values $y^{(p)}$:

$$F(x_1, \ldots, x_n) = J\left(y^{(1)}, \ldots, y^{(P_0)}\right).$$

If it so happens that for some tuple $(x_1, \ldots, x_n)$, the dependence on x_{n+1} is negligible, i.e., if for all possible values of the extra variable x_{n+1}, we have $f(x_1, \ldots, x_{n+1}) = f(x_1, \ldots, x_n, \overline{x}_{n+1})$, where by $\overline{x}_{n+1}$ denotes a "typical" (e.g., average) value of x_{n+1}, then we will have $y^{(1)} = \cdots = y^{(P_0)}$. In this case, the best fit is attained when $F(x_1, \ldots, x_n)$ is equal to all these values, i.e., when $F(x_1, \ldots, x_n) = f(x_1, \ldots, x_n, \overline{x}_{n+1})$.

It is therefore reasonable to expand the dependence J of the optimal value $F(x_1, \ldots, x_n)$ on the values $y^{(p)}$ in Taylor series around the point

$$\left(y^{(1)}, \ldots, y^{(P_0)}\right) = (f(x_1, \ldots, x_n, \overline{x}_{n+1}), \ldots, f(x_1, \ldots, x_n, \overline{x}_{n+1})).$$

This way, we get

$$F(x_1, \ldots, x_n) = f(x_1, \ldots, x_n, \overline{x}_{n+1}) + \sum_{1 \le p \le P_0} J_p \cdot \Delta y^{(p)} +$$

$$\sum_{1 \le p \le p' \le P_0} J_{p,p'} \cdot \Delta y^{(p)} \cdot \Delta y^{(p')} +$$

$$\sum_{1 \le p \le p' \le p'' \le P_0} J_{p,p',p''} \cdot \Delta y^{(p)} \cdot \Delta y^{(p')} \cdot \Delta y^{(p'')}, \tag{4}$$

where we denoted

$$\Delta y^{(p)} \stackrel{\text{def}}{=} y^{(p'')} - f(x_1, \ldots, x_n, \overline{x}_{n+1}).$$

Since we are interested only in the terms up to the third order—as we have found out, higher order terms are ignored anyway—it is sufficient to only consider terms up to this order in the expression (5).

Here, due to (4), we have

$$y^{(p)} - f(x_1, \ldots, x_n, \overline{x}_{n+1}) = f\left(x_1, \ldots, x_n, x_{n+1}^{(p)}\right) - f(x_1, \ldots, x_n, \overline{x}_{n+1}).$$

In general, substituting the explicit expression (3) for the function $f(x_1, \ldots, x_n, x_{n+1})$ into the expression for the difference

$$f(x_1, \ldots, x_n, x_{n+1}) - f(x_1, \ldots, x_n, \overline{x}_{n+1}),$$

and taking into account that terms not depending on x_{n+1} cancel each other, we conclude that

$$f(x_1, \ldots, x_n, x_{n+1}) - f(x_1, \ldots, x_n, \overline{x}_{n+1}) = a_{n+1} \cdot (x_{n+1} - \overline{x}_{n+1}) +$$

$$\sum_{1\le i\le n} a_{i,n+1}\cdot x_i\cdot(x_{n+1}-\overline{x}_{n+1})+a_{n+1,n+1}\cdot\left(x_{n+1}^2-(\overline{x}_{n+1})^2\right)+$$

$$\sum_{1\le i\le j\le n} a_{i,j,n+1}\cdot x_i\cdot x_j\cdot(x_{n+1}-\overline{x}_{n+1})+\sum_{1\le i\le n} a_{i,n+1,n+1}\cdot x_i\cdot\left(x_{n+1}^2-(\overline{x}_{n+1})^2\right)+$$

$$a_{n+1,n+1,n+1}\cdot\left(x_{n+1}^3-(\overline{x}_{n+1})^3\right). \tag{5}$$

In particular, in the case when $x_{n+1}=x_{n+1}^{(p)}$, we get

$$y^{(p)}-f(x_1,\ldots,x_n,\overline{x}_{n+1})=a_{n+1}\cdot\left(x_{n+1}^{(p)}-\overline{x}_{n+1}\right)+$$

$$\sum_{1\le i\le n} a_{i,n+1}\cdot x_i\cdot\left(x_{n+1}^{(p)}-\overline{x}_{n+1}\right)+a_{n+1,n+1}\cdot\left(\left(x_{n+1}^{(p)}\right)^2-(\overline{x}_{n+1})^2\right)+$$

$$\sum_{1\le i\le j\le n} a_{i,j,n+1}\cdot x_i\cdot x_j\cdot\left(x_{n+1}^{(p)}-\overline{x}_{n+1}\right)+$$

$$\sum_{1\le i\le n} a_{i,n+1,n+1}\cdot x_i\cdot\left(\left(x_{n+1}^{(p)}\right)^2-(\overline{x}_{n+1})^2\right)+$$

$$a_{n+1,n+1,n+1}\cdot\left(\left(x_{n+1}^{(p)}\right)^3-(\overline{x}_{n+1})^3\right). \tag{6}$$

We can see that in this expression, all the terms are proportional to the difference $x_{n+1}^{(p)}-\overline{x}_{n+1}$. Thus, terms in (5) which are quadratic or of third order with respect to the difference $y^{(p)}-f(x_1,\ldots,x_n,\overline{x}_{n+1})$ are also proportional to the difference $x_{n+1}^{(p)}-\overline{x}_{n+1}$. Thus, since we are only interested in terms which are at most cubic in terms of all the variables $x_1,\ldots,x_n,x_{n+1}$, and each of these terms contains some value of x_{n+1}, all these terms are at most quadratic in terms of $x_1,\ldots,x_n$. So, from the formula (5), we conclude that the difference $F(x_1,\ldots,x_n)-f(x_1,\ldots,x_n,\overline{x}_{n+1})$ is a quadratic function of $x_1,\ldots,x_n$, i.e., that

$$F(x_1,\ldots,x_n)-f(x_1,\ldots,x_n,\overline{x}_{n+1})=b_0+\sum_{1\le i\le n} b_i\cdot x_i+\sum_{1\le i\le j\le n} b_{i,j}\cdot x_i\cdot x_j \tag{7}$$

for some coefficients b_0, b_i, and $b_{i,j}$. From the formulas (6) and (8), we conclude that the difference

$$\Delta f(x_1,\ldots,x_{n+1}\stackrel{\text{def}}{=}f(x_1,\ldots,x_n,x_{n+1})-F(x_1,\ldots,x_n)=$$

$$(f(x_1,\ldots,x_n,x_{n+1})-f(x_1,\ldots,x_n,\overline{x}_{n+1}))-(F(x_1,\ldots,x_n)-f(x_1,\ldots,x_n,\overline{x}_{n+1}))$$

has the form

$$\Delta f(x_1, \ldots, x_n, x_{n+1}) = a_{n+1} \cdot (x_{n+1} - \overline{x}_{n+1}) + \sum_{1 \le i \le n} a_{i,n+1} \cdot x_i \cdot (x_{n+1} - \overline{x}_{n+1}) +$$

$$a_{n+1,n+1} \cdot \left(x_{n+1}^2 - (\overline{x}_{n+1})^2\right) + \sum_{1 \le i \le j \le n} a_{i,j,n+1} \cdot x_i \cdot x_j \cdot (x_{n+1} - \overline{x}_{n+1}) +$$

$$\sum_{1 \le i \le n} a_{i,n+1,n+1} \cdot x_i \cdot \left(x_{n+1}^2 - (\overline{x}_{n+1})^2\right) + a_{n+1,n+1,n+1} \cdot \left(x_{n+1}^3 - (\overline{x}_{n+1})^3\right) -$$

$$b_0 - \sum_{1 \le i \le n} b_i \cdot x_i - \sum_{1 \le i \le j \le n} b_{i,j} \cdot x_i \cdot x_j. \tag{8}$$

This expression contains the following unknowns:

- one parameter a_{n+1},
- n parameters $a_{i,n+1}$,
- one parameter $a_{n+1,n+1}$,
- $\dfrac{n \cdot (n+1)}{2}$ parameters $a_{i,j,n+1}$,
- n parameters $a_{i,n+1,n+1}$,
- one parameter $a_{n+1,n+1,n+1}$,
- one parameter b_0,
- n parameters b_i, and
- $\dfrac{n \cdot (n+1)}{2}$ parameters $b_{i,j}$.

to the total of

$$1 + n + 1 + \frac{n \cdot (n+1)}{2} + n + 1 + 1 + \frac{n \cdot (n+1)}{2} = n^2 + 3 \cdot n + 2 \text{ parameters.}$$

Let us recall that in a shallow network with $n+1$ inputs and $n+1$ neurons in the hidden layer, we have $(n+2)^2 = n^2 + 4 \cdot n + 4$ parameters, and

$$n^2 + 4 \cdot n + 4 > n^2 + 3 \cdot n + 2.$$

Thus, we can fit all the new cubic terms if we train a shallow neural network to recognize the difference between:

- the actual values $y^{(p)}$ of the output, and
- the values $F\left(x_1^{(p)}, \ldots, x_n^{(p)}\right)$ produced by the pre-trained neural network (that does not take the new quantity x_{n+1} into account).

Then, the result of this training should be simply added to the result of pre-trained neural network.

Discussion. Specifically, the output of the shallow network should be added, to the pre-trained neural network, as an extra neuron in the penultimate layer, with weight 1 from this neuron to the final output neuron of the whole neural network. (Strictly speaking, this will make the resulting network residual.)

This will save time, since:

- training time is, crudely speaking, proportional to the number of parameters that we need to determine, and
- in our case, we decrease this number from n^3—for the straightforward re-training—to the value n^2 needed to train the shallow network.

This way, we will cover all linear, quadratic, and cubic terms in the dependence of the output on all $n+1$ quantities $x_1, \ldots, x_n, x_{n+1}$. This will not cover 4th order terms—but these terms, as we have mentioned, are not covered by deep learning anyway.

Let us describe our proposal in precise terms.

3 Resulting Proposal

Formulation of the problem: reminder.

- *What we have*: We have a neural network pre-trained to describe the dependence of a quantity y on quantities $x_1, \ldots, x_n$. We will denote the result of applying this trained network to the inputs $x_1, \ldots, x_n$ by $F(x_1, \ldots, x_n)$.
- *What we want*: We would like to modify this network, so that it will take into account dependence on an additional quantity x_{n+1} as well.

A straightforward way to do it is to add one more input x_{n+1}, and to re-train all the weights of the whole original network by using all the patterns that contain the value of this input. The problem is that this would take a long time, so a question is: can we do it faster?

Specific recommendation. We can speed up the process if we do the following:

- First, we train a shallow neural network with n intermediate neurons to describe the dependence of the difference $Y^{(p)} = y^{(p)} - F\left(x_1^{(p)}, \ldots, x_n^{(p)}\right)$ on all the inputs $x_1^{(p)}, \ldots, x_n^{(p)}, x_{n+1}^{(p)}$ for which we know the value of the extra variable. Let us denote the result of applying the resulting trained shallow neural network to the inputs $x_1, \ldots, x_n, x_{n+1}$ by $S(x_1, \ldots, x_n, x_{n+1})$.
- Then, as a re-trained neural network, we take the network that computes the value $F(x_1, \ldots, x_n) + S(x_1, \ldots, x_n, x_{n+1})$. For this purpose, we:
  - add the output neuron of the shallow network to the penultimate layer of the original network, and

- set the weight connecting this new neuron to the output layer of the original network to 1.

Further discussion. The above recommendation is the one that is formally justified by our analysis. However, less formally, we can say that the above specific scheme encourages us to use similar simplified re-training schemes. Indeed, our main idea was that:

- since we need to determine n^2 new parameters,
- we should not re-train all n^3 weights, it is most probably sufficient to only change n^2 weights.

How can we find a part of the network that contains exactly n^2 weights? This is easy: as we have mentioned earlier, each layer of a deep neural network contains n^2 weights. Thus, a reasonable idea is to re-train only one layer, while leaving all other weights unchanged ("frozen").

Which layer should we choose? We need to involve an additional input x_{n+1}.

Resulting idea. So, a reasonable idea is to train only the weights of the first layer—the layer that (directly) processes the inputs—while keeping all other weights unchanged. This general idea is what we tried.

4 Experiments

Our preliminary results show that this faster training indeed leads to results which are as accurate as the full training. For this testing, we use the following example from pavement engineering, where the goal is to estimate the value of the failure function—that gauges the stability f of an untreated pavement layer [1, 2]. Traditionally, this function is estimated based on four inputs: the three principal stresses σ_1, σ_2, and σ_3, and the angle of internal friction φ. The algorithm for estimating f based these four values consists of the following steps:

- First, we compute the first two invariants of the stress tensor:

$$I_1 = \sigma_1 + \sigma_2 + \sigma_3, \text{ and } J_2 = \frac{1}{6} \cdot \left[(\sigma_1 - \sigma_2)^2 + (\sigma_1 - \sigma_3)^2 + (\sigma_2 - \sigma_3)^2\right].$$

- Based on the inputs σ_i and on the first invariant I_1, we compute the value of the third invariant:

$$J_3 = \left[\sigma_1 - \frac{I_1}{3}\right] \cdot \left[\sigma_2 - \frac{I_1}{3}\right] \left[\sigma_3 - \frac{I_1}{3}\right].$$

- Then, we compute the angle θ based on the formula

$$\cos(3 \cdot \theta) = \frac{3 \cdot \sqrt{3}}{2} \cdot \frac{J_3}{J_2^{3/2}}.$$

- Finally, we estimate the value of the failure function f as

$$f_0 = \frac{I_3}{2} \cdot \sin(\varphi) + \sqrt{J_2} \cdot \sin\left(\theta + \frac{\pi}{2}\right) + \frac{\sqrt{J_3}}{3} \cdot \cos\left(\theta + \frac{\pi}{2}\right) \cdot \sin(\varphi). \quad (9)$$

A slightly more accurate estimate can be obtained if we take into account the value of an additional quantity: cohesion c. The corresponding formula has the form

$$f = f_0 - c \cdot \cos(\varphi). \quad (10)$$

To test our idea, we formed a large number of randomly selected tuples $(\sigma_1^{(p)}, \sigma_2^{(p)}, \sigma_3^{(p)}, \varphi^{(p)}, c^{(p)})$. For each of these tuples, we used the formula (10) to compute the corresponding value $f^{(p)}$ of the failure function. In line with the above idea:

- **Task 1 (Without c)**: We trained the neural network on the patterns $(\sigma_1^{(p)}, \sigma_2^{(p)}, \sigma_3^{(p)}, \varphi^{(p)}, y^{(p)})$ that only included the values of the four original inputs.
- **Task 2 (1 layer training)**: We added an extra input corresponding to the extra variable c, froze the weights of all the layers except for the first one—so that only weights in the first layer that connects the input variable c to the next layer are changed—and trained these weights on the full tuples $(\sigma_1^{(p)}, \sigma_2^{(p)}, \sigma_3^{(p)}, \varphi^{(p)}, c^{(p)}, y^{(p)})$, including the values of the extra variable c.
- **Task 3 (full network training)**: For comparison, we performed a similar re-training without freezing, when the weights in all the layers were allowed to change during training.

For each of the three training tasks, we used a feedforward neural network with four internal layers $[128 \times 64 \times 32 \times 16]$ using the Adam optimizer with a learning rate of 0.001, a batch size of 64, and mean squared error as the loss function. To freeze the weights, we customized PyTorch library to define layers, linear operators, and ReLU non-linear activation function. We used `torch.no_grad()` option to froze every weights and layers except ones that connect the new variable to the next layer.

Table 1 shows the results of our experiments. For each training, we ran 2,000 iterations and recorded the results after 500, 1000, 1500, and 2000 iterations (first row). The first part of table (rows 2 to 5) shows the results without having the variable c (Task 1), the second part of table (rows 6 to 9) shows the results with adaptive training of first layer after adding the variable c, and the third part of table (rows 10 to 13) shows the results of full network training after adding the variable c. We reported the loss, the accuracy that measure how many predictions were within the unit distance of ground truth, the R^2 metric that measures the correlations between predictions and ground truth, and the computation times of training in seconds. We fixed the seed to minimize the randomness of functional outcomes (loss, accuracy, and R^2). To control the noise of environment for time measurements, we used an isolated server computer with Intel Xeon CPU with 2 vCPUs (virtual CPUs) and 13GB of RAM.

Table 1 Experimental results

# of iterations	500	1,000	1,500	2,000
Loss (Without c)	4.2	0.2	0.2	0.2
Accuracy (Without c)	47.0%	97.0%	98.0%	99.0%
R-Regression (Without c)	0.82	0.99	0.99	0.99
Training time (Without c)	147(s)	314(s)	436(s)	600(s)
Loss (1-layer training)	0.4	0.6	0.9	0.1
Accuracy (1-layer training)	85.6%	83.0%	81.4%	99.9%
R-Regression (1-layer training)	0.98	0.98	0.97	0.99
Training time (1-layer training)	148(s)	292(s)	405(s)	552(s)
Loss (Full training)	2.8	1.3	0.2	0.4
Accuracy (Full training)	53.0%	72.2%	97.0%	86.0%
R-Regression (Full training)	0.91	0.95	0.99	0.98
Training time (Full training)	164(s)	300(s)	410(s)	555(s)

We observed that the proposed approach of adaptive training of first layer achieves better computation times where the differences in the computation times are reduced as the number of iterations increase. Within 2,000 iterations, the three approaches (without c, first layer training, and full training) achieve 99.0, 99.9, and 97% accuracy, respectively. Since there are fluctuations in the accuracy, one can use an early stop in training the network, and the highest accuracy shows the best results that can be achieved in each approach.

Acknowledgements This work was supported in part by the National Science Foundation grants 1623190 (A Model of Change for Preparing a New Generation for Professional Practice in Computer Science), HRD-1834620 and HRD-2034030 (CAHSI Includes), EAR-2225395, and by the AT&T Fellowship in Information Technology.

It was also supported by the program of the development of the Scientific-Educational Mathematical Center of Volga Federal District No. 075-02-2020-1478, and by a grant from the Hungarian National Research, Development and Innovation Office (NRDI).

The authors are thankful to Reza Ashtiani and Edgar Rodriguez for their help with the pavement engineering example.

References

1. Ashtiani, R.S.: Anisotropic Characterization and Performance Prediction of Chemically and Hydraulically Bounded Pavement Foundations, Doctoral dissertation, Texas A&M University. Available electronically from: https://hdl.handle.net/1969.1/ETD-TAMU-2009-08-7103
2. Ashtiani, R.S., Luo, R., Lytton, R.L.: Performance prediction and moisture susceptibility of anisotropic pavement foundations. In: Huang, B., Tutumluer, E., Al-Qadi, I.L., Prozzi, J., Shu, X. (eds.) Paving Materials and Pavement Analysis: Proceedings of the GeoShanghai 2010 Conference, pp. 327–334. American Society of Civil Engineers (ASCE), Reston, Virginia (2010)
3. Bishop, C.M.: Pattern Recognition and Machine Learning. Springer, New York (2006)
4. Feynman, R., Leighton, R., Sands, M.: The Feynman Lectures on Physics. Addison Wesley, Boston, Massachusetts (2005)
5. Goodfellow, I., Bengio, Y., Courville, A.: Deep Learning. MIT Press, Cambridge, Massachusetts (2016)
6. Holguin, S., Kreinovich, V.: Why residual neural networks. In: Ceberio, M., Kreinovich, V. (eds.) Decision Making Under Uncertainty and Constraints: A Why-Book, pp. 117–120. Springer, Cham, Switzerland (2023)
7. Thorne, K.S., Blandford, R.D.: Modern Classical Physics: Optics, Fluids, Plasmas, Elasticity, Relativity, and Statistical Physics. Princeton University Press, Princeton, New Jersey (2021)

Towards a Psychologically Natural Relation Between Colors and Fuzzy Degrees

Victor L. Timchenko, Yuriy P. Kondratenko, Olga Kosheleva, Vladik Kreinovich, and Nguyen Hoang Phuong

Abstract A natural way to speed up computations—in particular, computations that involve processing fuzzy data—is to use the fastest possible communication medium: light. Light consists of components of different color. So, if we use optical color computations to process fuzzy data, we need to associate fuzzy degrees with colors. One of the main features—and of the main advantages—of fuzzy technique is that the corresponding data has intuitive natural meaning: this data comes from words from natural language. It is desirable to preserve this naturalness as much as possible. In particular, it is desirable to come up with a natural relation between colors and fuzzy degrees. In this paper, we show that there is exactly one such natural relation, and we describe this relation.

V. L. Timchenko
Admiral Makarov National University of Shipbuilding, 9 Heroes of Ukraine Avenue, Mykolaiv 054025, Ukraine

Y. P. Kondratenko
Petro Mohyla Black Sea National University, 10, 68th Desantnykiv Str., Mykolaiv 054003, Ukraine

Institute of Artificial Intelligence Problems, Ministry of Education and Science (MES) of Ukraine and National Academy of Sciences (NAS) of Ukraine, Kyiv 01001, Ukraine

O. Kosheleva
Department of Teacher Education, University of Texas at El Paso, 500 W. University El Paso, El Paso, TX 79968, USA
e-mail: olgak@utep.edu

V. Kreinovich (✉)
Department of Computer Science, University of Texas at El Paso, 500 W. University El Paso, El Paso, TX 79968, USA
e-mail: vladik@utep.edu

N. Hoang Phuong
Artificial Intelligence Division, Information Technology Faculty, Thang Long University, Nghiem Xuan Yem Road, Hanoi, Hoang Mai District, Vietnam

N. Hoang Phuong et al. (eds.), *Machine Learning and Other Soft Computing Techniques: Biomedical and Related Applications*, Studies in Systems, Decision and Control 543, https://doi.org/10.1007/978-3-031-63929-6_5

1 Formulation of the Problem

Need for processing fuzzy information. In many real-life situations—be it driving a car or treating a patient—humans provide a good quality control, often control which is of better quality than what current automatic controllers can provide. It is therefore desirable to incorporate the knowledge and experience of these human controllers into the automatic systems, to make these systems more effective.

Such incorporation is not easy, and one of the main reasons why it is not easy is that human experts often describe their control strategies not in precise easy-to-program mathematical form, but by using imprecise ("fuzzy") words from natural language such as "small". To transform this knowledge into precise computer-understandable terms, Lotfi Zadeh invented a special technique that he called *fuzzy*; see, e.g., [2, 6, 7, 9, 10, 23].

In this technique, to describe an imprecise property like "small", we ask the expert to assign, to each possible value x of the corresponding quantity, a degree—from the interval [0, 1]—describing to what extend this value has the given property (e.g., is small).

Need for discrete fuzzy degrees. From the purely mathematical viewpoint, there are infinitely many degrees in the interval [0, 1], so nothing should prevent experts to assign degrees like 0.2134 or 0.6778. However, in practice, people cannot meaningfully distinguish between many degrees. According to psychology, they can distinguish between 7 plus minus 2 different degrees; see, e.g., [8, 12]. For most people, this number is 7, so it makes sense to use seven different degrees.

So, in addition to absolute "yes", absolute "no", and a neutral situation when we have exactly as many arguments for "yes" as for "no", we should have four degrees corresponding to some degree of confidence in a statement or it is negation. Thus, we arrive at the following natural list of degrees:

No	Very probably No	Probably No	Neutral	Probably Yes	Very probably Yes	Yes

Need for fast computations. In many practical problems, especially in problems related to real-time control, it is desirable to perform computations as fast as possible.

Need for optical computing. One of the factors that limits the computation speed is the speed with which signals pass from one part of the processing unit to another. It is therefore desirable to make sure that this signal transfer occurs at the highest possible speed—and according to physics, the fastest possible speed is the speed of light; see, e.g., [5, 14]. Thus, to speed up computations, it is desirable to use, in computations, electromagnetic waves—in particular, visible light. Such computations are known as *optical computing*.

Need for color optical computing. It is well known that every optical signal can be represented as a combination of several signals of pure color. This fact that discovered by Newton, who showed that each light can be decomposed into pure colors—and then these pure colors can be assembled back, into the original signal. Thus, a natural idea is to use signals of different color. Such computations are known as *color optical computing*.

In particular, it is desirable to use color optical computing to process fuzzy data; see, e.g., [15–22].

Historical comment. From the mathematical and signal processing viewpoints, Newton's discovery about colors was the first empirical proof of what is now known as Fourier techniques (see, e.g., [11, 13])—whose main idea is that each signal can be presented as a combination of sinusoids with different frequency (and pure colors are exactly signals of fixed frequency).

Problem: it is desirable to have a natural relation between colors and fuzzy degrees. To use color optical computing for processing fuzzy data, it is desirable to assign different colors to different fuzzy degrees. Fuzzy techniques process intuitively clear notions like "small". It would be advantageous to extend this naturalness to data processing as well, i.e., to make the relation between colors and fuzzy degrees natural.

What we do in this paper. In this paper, we show how to come up with such a natural relation between colors and fuzzy degrees.

2 Towards the Desired Natural Relation

What colors should we consider? It is known that in our color perception, we have three basic colors—red (R), blue (B), and green (G). From our viewpoint, every other color can be represented as a combination of these three basic colors:

- a combination of red and green (R + G) leads to yellow (Y),
- a combination of red and blue (R + B) leads to magenta (M),
- a combination of green and blue (G + B) leads to cyan (C), and
- a combination of all three colors (R + G + B) leads to white (W).

Also, the absence of colors corresponds to black (Blc). These are the colors that we need to assign to different fuzzy degrees.

What is a natural analog of "neutral"? The most easy thing is to assign a color to the degree "neural". Indeed, in this degree, the expert has exactly as many argument for as arguments again. In other words, in such situation, the expert cannot decide, i.e., in effect, has no contribution to the decision process. It is therefore natural to associate this no-information degree with no-information color—i.e., with the color Black.

What remains to be done is to associate appropriate colors with the remaining six degrees:

No	Very probably no	Probably no	Probably yes	Very probably yes	Yes
?	?	?	?	?	?

What is known. Psychologists have studied the relation between colors and decision making, and they found an association of some colors with decisions; see, e.g., [1, 3, 4]. (One has to be cautious since this research is mostly based on US customers, but still, these associations are largely universal.)

Let us sort this information in the order in which these colors appear in the spectrum:

- The color Yellow is associated with Inexpensive and Low Quality.
- The color Green is associated with Good Taste.
- The color Blue is associated with High Quality and Reliability.

Also, the color Black is associated with Expensive.

How can we translate these associations into fuzzy degrees?

- Let us start with Yellow. Low Quality is not good. However, because of Inexpensive, it is not a clear No: sometimes, we buy low quality things because we cannot afford things of better quality. We are not happy when we do it, so a natural idea is to assign, to this color, the degree "very probably no".
- Good Taste is a positive characteristic, but it is only one—and often, not the most important characteristic. More important is how usable it is, taste is secondary. It is therefore natural to associate, to this color, the degree "probably yes".
- Blue has only positive characteristics, both are important, so let us associate it with "yes".
- What about Black? The fact that something is Expensive does not mean that we should buy it—only snobs buy things just because they are expensive. It also does not mean that we should *not* buy it: expensive is a price that we pay for high quality. In other words, from the viewpoint of decision making, the very fact that something is expensive does not help to decide whether we but it or not, it is neutral—which is in perfect agreement with our previous association of Black with "neutral".

Let us summarize the above arguments by placing the corresponding colors into our table:

No	Very probably no	Probably no	Probably yes	Very probably yes	Yes
?	R + G	?	G	?	B

How to extend this assignment to other colors: a natural idea. We need to extend the above psychologically natural relation between colors and fuzzy degrees to the remaining degrees. A natural idea is that the degree corresponding to the mixture of two colors should be located between the degrees corresponding to these two colors. So, we come up with the following conclusions.

According to the above idea, Y = R + G should be between R and G. In our table, the degree corresponding to G is to the right of the degree corresponding to Y. So, the degree corresponding to R should be to the left of the degree "most probably no" corresponding to Y. The only degree which is to the left of "very probably no" is "no". So, we assign, to Red, the degree "no":

No	Very probably no	Probably no	Probably yes	Very probably yes	Yes
R	R + G	?	G	?	B

Similar, the degree corresponding to G + B should be in between the degrees corresponding to G and B, i.e., in between the degrees "probably yes" and "yes". There is one degree in between: "very probably yes". So, we assign this degree to the color G + B:

No	Very probably no	Probably no	Probably yes	Very probably yes	Yes
R	R + G	?	G	G + B	B

The only remained un-assigned degree is "probably no", and the only remaining un-assigned color is R + B, so we naturally match them. (This matching, by the way, is in perfect accordance with our idea, since this degree is in between degrees "no" and "yes" corresponding to Red and Blue.) Thus, we arrive at the following conclusion.

Conclusion. The following is the natural association of colors and fuzzy degrees:

No	Very probably no	Probably no	Probably yes	Very probably yes	Yes
R	R + G	R + B	G	G + B	B

To make this relation clearer, let us reformulate it with full names of the resulting colors:

Color	Structure	Degree
Red (R)	R	No
Yellow (Y)	R + G	Most probably no
Magenta (M)	R + B	Probably no
Black		Neutral
Green (G)	G	Very probably yes
Cyan (C)	G + B	Probably yes
Blue (B)	B	Yes

3 Discussion

Can we make this natural relation less subjective? Our relation relies on the results of empirical psychological studies. A natural question is: can we make our conclusions less subjective, based only on some objective characteristics—so that we will not need to rely on these empirical results? In this section, we will try to do it.

What is a natural objective characteristic of a color? As we have mentioned, a color is described by its frequency. We are talking about using color to communicate.

From this viewpoint, the higher the frequency, the most information we can transfer by using signals of this color. So, the larger the frequency, the more positive our attitude to different colors. Let us show how this idea helps.

Let us start with the ordering of three basic colors. Among the three basic colors Red, Green, and Blue, Red corresponds to the smallest frequency, Green to the intermediate one, and Blue to the largest one, so we have:

$$\mathrm{R} < \mathrm{G} < \mathrm{B}.$$

Let us extend this order to combinations of basic colors. It is natural to assume—as we did before—that the quality of a combination of two colors should be in between the corresponding colors, i.e., that

$$\mathrm{R} < \mathrm{R} + \mathrm{G} < \mathrm{G} < \mathrm{G} + \mathrm{B} < \mathrm{B}$$

and

$$\mathrm{R} < \mathrm{R} + \mathrm{B} < \mathrm{B}.$$

It is also reasonable to require that if we replace one color in a combination with a better color, the quality of the combination improves. As a result, we get

$$\mathrm{R} + \mathrm{G} < \mathrm{R} + \mathrm{B} < \mathrm{G} + \mathrm{B}.$$

Resulting order. By combining the above comparisons, we end up with the following order:

$$\mathrm{R} < \mathrm{R} + \mathrm{G} < \mathrm{G}, \mathrm{R} + \mathrm{B} < \mathrm{G} + \mathrm{B} < \mathrm{B}. \tag{1}$$

This is *almost* what we got by using the empirical data, but not exactly: by using the general ideas, we cannot decide which is better: Green (G) or Magenta (M = R + B).

Why cannot we get the full order this way? It is easy to see why we cannot get the full order based only on the objective characteristics. Indeed, the original situation remains the same if we reverse the order, i.e., if we:

- swap R and B, while keeping G intact, and
- replace each $<$ with $>$.

One can easily see that the resulting relations (1) do not change under this transformation. However:

- we cannot have R + B $<$ G, since this transformation will transform it into its opposite R + B $>$ G, and these two inequalities cannot be both true;
- similarly, we cannot have G $<$ R + B, since this transformation will transform it into its opposite G $>$ R + B, and these two inequalities cannot be both true.

Comment. This may be related to fact that, in contrast to Green and to many other colors, magenta (R + B) is *not* a real color—in the sense that there is no frequency corresponding to magenta. The only way to get a perception of magenta is to mix several colors. This is different from other combinations, such as Yellow:

- yes, in the Red-Green-Blue scheme, Yellow is a combination of two colors,
- but there is also a wavelength range 565–590 THz corresponding to Yellow.

Similarly, there is a wavelength range 600–620 THz corresponding to Cyan (which we represented as G + B).

But maybe we can get the full order this way? What if we take into account not only the *order* between the frequencies corresponding to three basic colors, but also the *numerical ranges* of these frequencies:

- 440–480 THz for Red, centered at 440,
- 530–600 THz for Green, centered at 565, and
- 620–670 THz for Blue, centered at 645,

we can see that the central point (565) of the Green frequency range is much further to the right than the average $(440 + 645)/2 = 542.5$ of the central points of the frequency ranges corresponding to Red and Blue. From this viewpoint, it is not surprising that G feels better that R + G.

Acknowledgements This work was supported in part by the National Science Foundation grants 1623190 (A Model of Change for Preparing a New Generation for Professional Practice in Computer Science), HRD-1834620 and HRD-2034030 (CAHSI Includes), EAR-2225395 (Center for Collective Impact in Earthquake Science C-CIES), and by the AT&T Fellowship in Information Technology.
It was also supported by a grant from the Hungarian National Research, Development and Innovation Office (NRDI).

References

1. Aslam, M.M.: Are you selling the right colour? a cross-cultural review of colour as a marketing cue. J. Market. Commun. **12**(1), 15–30 (2006)
2. Belohlavek, R., Dauben, J.W., Klir, G.J.: Fuzzy Logic and Mathematics: A Historical Perspective. Oxford University Press, New York (2017)
3. Bottomley, P.A., Doyle, J.R.: The interactive effects of colors and products on perceptions of brand logo appropriateness. Market. Theory **6**(1), 63–83 (2006)
4. Color Psychology, Wikipedia, https://en.wikipedia.org/wiki/Color_Psychology. Accessed 1 Aug 2023
5. Feynman, R., Leighton, R., Sands, M.: The Feynman Lectures on Physics. Addison Wesley, Boston, Massachusetts (2005)
6. Klir, G., Yuan, B.: Fuzzy Sets and Fuzzy Logic. Prentice Hall, Upper Saddle River, New Jersey (1995)
7. Mendel, J.M.: Uncertain Rule-Based Fuzzy Systems: Introduction and New Directions. Springer, Cham, Switzerland (2017)
8. Miller, G.A.: The magical number seven plus or minus two: some limits on our capacity for processing information. Psychol. Rev. **63**(2), 81–97 (1956)

9. Nguyen, H.T., Walker, C.L., Walker, E.A.: A First Course in Fuzzy Logic. Chapman and Hall/CRC, Boca Raton, Florida (2019)
10. Novák, V., Perfilieva, I., Močkoř, J.: Mathematical Principles of Fuzzy Logic. Kluwer, Boston, Dordrecht (1999)
11. Osgood, B.G.: Lectures on the Fourier Transform and Its Applications. American Mathematical Society, Providence, Rhode Island (2019)
12. Reed, S.K.: Cognition: Theories and Application. SAGE Publications, Thousand Oaks, California (2022)
13. Stein, E.M., Shakarchi, R.: Fourier Analysis: An Introduction. Princeton University Press, Princeton, New Jersey (2003)
14. Thorne, K.S., Blandford, R.D.: Modern Classical Physics: Optics, Fluids, Plasmas, Elasticity, Relativity, and Statistical Physics. Princeton University Press, Princeton, New Jersey (2021)
15. Timchenko, V., Kondratenko, Y., Kreinovich, V.: Efficient optical approach to fuzzy data processing based on colors and light filter. Int. J. Problems Control Inf. **52**(4), 89–105 (2022)
16. Timchenko, V., Kondratenko, Y., Kreinovich, V.: Decision support system for the safety of ship navigation based on optical color logic gates. In: Proceedings of the IX International Conference Information Technology and Implementation IT & I-2022, CEUR Workshop Proceedings, Vol. 3347, pp. 42–52. Kyiv, Ukraine. http://ceur-ws.org/
17. Timchenko, V., Kondratenko, Y., Kreinovich, V.: Implementation of optical logic gates based on color filters. In: Hu, Z., et al. (eds.) Proceedings of the he 6th International Conference on Computer Science, Engineering and Education Applications ICCSEEA 2023, Lecture Notes on Data Engineering and Communications Technologies, Vol. 181. Warsaw, Poland (2023). https://doi.org/10.1007/978-3-031-36118-0_12
18. Timchenko, V.L., Kondratenko, Y.P., Kreinovich, V.: The architecture of optical logical coloroid with fuzzy computing. In: Proceedings of the 4th International Workshop on Intelligent Information Technologies & Systems of Information Security IntelITSIS'2023, CEUR Workshop Proceedings, 2023, Vol. 3373, pp. 638–648. Khmelnitsky, Ukraine (2023)
19. Timchenko, V.L., Kondratenko, Y.P., Kreinovich, V., Kosheleva, O.: Natural color interpretation of interval-valued fuzzy degrees. In: Proceedings of the 20th World Congress of the International Fuzzy Systems Association IFSA'2023. Daegu, South Korea (2023)
20. Timchenko, V., Kondratenko, Y., Kozlov, O., Kreinovich, V.: Fuzzy color computing based on optical logical architecture. Proceedings of the International Conference on Intelligent and Fuzzy Systems INFUS'2023. Istanbul, Turkey (2023)
21. Timchenko, V.L., Kondratenko, Y.P., Kreinovich, V.: Interval-valued and set-valued extensions of discrete fuzzy logics, belnap logic, and color optical computing. In: Proceedings of the Biennial Conference of the European Society for Fuzzy Logic and Technology EUSFLAT'2023. Palma de Mallorca, Spain (2023)
22. Timchenko, V.L., Kondratenko, Y.P., Kreinovich, V.: Why color optical computing. In: Hoang Phuong, N., Kreinovich, V. (eds.) Deep Learning and Other Soft Computing Techniques: Biomedical and Related Applications, Studies in Computational Intelligence, Vol. 1097, pp. 227–233. Springer, Cham, Switzerland (2023)
23. Zadeh, L.A.: Fuzzy sets. Inf. Control **8**, 338–353 (1965)

Algebraic Product Is the only "And-Like"-Operation for Which Normalized Intersection Is Associative: A Proof

Thierry Denœux and Vladik Kreinovich

Abstract For normalized fuzzy sets, intersection is, in general, not normalized. So, if we want to limit ourselves to normalized fuzzy sets, we need to normalize the intersection. It is known that for algebraic product, the normalized intersection is associative, and that for many other "and"-operations (t-norms), normalized intersection is not associative. In this paper, we prove that algebraic product is the only "and"-operation (and even the only "and-like" operation) for which normalized intersection is associative.

1 Formulation of the Problem

Fuzzy sets and normalized fuzzy sets: a brief reminder. A *fuzzy set* on a universal set X is a function $\mu(x)$ that assigns, to each element $x \in X$, a number from the interval [0, 1]; see, e.g., [1, 3–7]. Fuzzy sets were invented to describe imprecise ("fuzzy") natural-language properties such as "small"; the value $\mu(x)$ is then a degree to which, according to the user, the object x has the desired property (e.g., is small).

- The degree 1 means that x definitely has the property.
- The degree 0 means that x definitely does not have this property.
- Intermediate value $\mu(x)$ correspond to x having the property "to some degree".

Usually, an additional condition is imposed that the fuzzy set should be *normalized*, i.e., that $\sup\limits_{x \in X} \mu(x) = 1$.

T. Denœux
Université de Technologie de Compiègne, U.M.R. C.N.R.S. 6599 Heudiasyc, Centre de Recherches de Royallieu, B.P. 20529 F -60205, Compiègne Cedex, France
e-mail: Thierry.Denoeux@utc.fr

Institut Universitaire de France, Paris, France

V. Kreinovich (✉)
Department of Computer Science, University of Texas at El Paso, 500 W. University, El Paso, TX 79968, USA
e-mail: vladik@utep.edu

N. Hoang Phuong et al. (eds.), *Machine Learning and Other Soft Computing Techniques: Biomedical and Related Applications*, Studies in Systems, Decision and Control 543, https://doi.org/10.1007/978-3-031-63929-6_6

"And"-operations (t-norms): a brief reminder. In many cases, we know the degrees a and b to which the properties A and B are satisfied, and we want to use this information to estimate the degree to which the property "A and B" is satisfied. The function $f_\&(a, b)$ that generates the resulting estimate is known as an *"and"-operation*, or, for historical reason, a *t-norm*.

This function have several natural properties. For example, since $A \,\&\, B$ means the same as $B \,\&\, A$, the corresponding estimates for $A \,\&\, B$ and $B \,\&\, A$ should coincide, i.e., we should have $f_\&(a, b) = f_\&(b, a)$ (i.e., in mathematical terms, the "and"-operation should be commutative). Similarly, since $(A \,\&\, B) \,\&\, C$ means the same as $A \,\&\, (B \,\&\, C)$, we should always have $f_\&(f_\&(a, b), c) = f_\&(a, f_\&(b, c))$ (i.e., in mathematical terms, the "and"-operation should be associative).

Also, intuitively, if A is absolutely true, i.e., if $a = 1$, then our degree of confidence in $A \,\&\, B$ is equal to our degree of confidence in B. In precise terms, this means that we should have $f_\&(1, b) = b$ for all b. Due to commutativity, we also have $f_\&(a, 1) = a$ for all a.

Intersection and normalized intersection of fuzzy sets. In general, if we have two properties–as described by the sets S_1 and S_2 of all the elements that satisfy the corresponding property, then the set of all the objects that satisfy the first property *and* satisfy the second property is called the *intersection* $S_1 \cap S_2$ of the two sets. In the fuzzy case, by the meaning of the fuzzy set, the degrees to which an object x satisfies each of the properties are equal to $\mu_1(x)$ and $\mu_2(x)$. Thus, by the meaning of the "and"-operation, the degree to which the first property is satisfied *and* the second property is satisfied is equal to $f_\&(\mu_1(x), \mu_2(x))$ for an appropriate "and"-operation $f_\&(a, b)$. So, it is reasonable to define the intersection of two fuzzy sets as

$$\mu(x) \stackrel{\text{def}}{=} f_\&(\mu_1(x), \mu_2(x)). \tag{1}$$

The problem with this definition is that if we, as usual, limit ourselves to normalized sets, then the above definition may lead to a non-normalized set. For example, if some $a < 1$, on the universal set $X = \{1, 2\}$, we consider two functions

$$f_1(1) = a, \quad f_1(2) = 1, \quad f_2(1) = 1, \quad f_2(2) = a,$$

then, in view of the fact that $f_\&(a, 1) = f_\&(1, a) = a$, we get $\mu(1) = \mu(2) = a < 1$, so $\max(\mu(x)) = a < 1$. If want the intersection to be a normalized fuzzy set, we must *normalize* the expression (1), i.e., divide it by the supremum of this expression:

$$(\mu_1 \,\&\, \mu_2)(x) = \frac{f_\&(\mu_1(x), \mu_2(x))}{\sup\limits_{y \in X} f_\&(\mu_1(y), \mu_2(y))}. \tag{2}$$

It is reasonable to call the operation (2) *normalized intersection*.

Comment. The intersection is only defined when the denominator is not equal to 0.

When is normalized intersection associative? By definition (2), normalized intersection operation (2) is commutative. It is known that for algebraic product $f_{\&}(a, b) = a \cdot b$, operation (2) is associative [2]. For many other "and"-operations, normalized intersection is not associative. So, a natural conjecture emerged that the algebraic product is the only "and"-operation for which the normalized intersection is associative.

What we do in this paper. In this paper, we prove that the algebraic product is indeed the only "and"-operation for which the normalized intersection is associative. Moreover, we prove it not only for all "and"-operations, but also for more general binary operations that are not necessarily commutative or associative.

2 Main Result

Definition 1 By an *"and"-like* operation, we mean a function $t : [0, 1] \times [0, 1] \mapsto [0, 1]$ for which $t(a, 1) = t(1, a) = a$ for all $a \in [0, 1]$.

Comment. This definition is weaker than that of a t-norm. In particular, we do not require monotonicity, or even associativity.

Definition 2 For each "and"-like operation $t(a, b)$, the corresponding *normalized intersection* operation transforms two fuzzy sets $\mu_1(x)$ and $\mu_2(x)$ into a new fuzzy set

$$(\mu_1 \,\&\, \mu_2)(x) = \frac{t(\mu_1(x), \mu_2(x))}{\sup\limits_{y \in X} t(\mu_1(y), \mu_2(y))}. \tag{3}$$

Comment. The intersection is only defined when the denominator of the formula (3) is different from 0.

Proposition. *For each "and"-like operation t(a,b), the following two conditions are equivalent:*

- *t(a,b) is the algebraic product, i.e.,* $t(a, b) = a \cdot b$;
- *The normalized intersection operation corresponding to, t(a,v) is associative, i.e., for all possible normalized fuzzy sets* μ_1, μ_2, and μ_3:
  - *the fuzzy sets* $(\mu_1 \cap \mu_2) \cap \mu_3$ and $\mu_1 \cap (\mu_2 \cap \mu_3)$ *are either both defined or both undefined, and*
  - *if they are both defined, then* $(\mu_1 \cap \mu_2) \cap \mu_3 = \mu_1 \cap (\mu_2 \cap \mu_3)$.

Discussion. In this formulation, we only use two properties out of many usual properties of the "and"-operation (t-norm):

- that for every a, we have $t(1, a) = a$, and
- that for every a, we have $t(a, 1) = a$.

From the mathematical viewpoint, a natural question is: can we go even further, keep only one of these properties, and still keep our result? The following two simple examples show that both above properties are needed, one is not enough to conclude that $t(a, b) = a \cdot b$:

- The function $t(a, b) = b$ has the property that $t(1, a) = a$ for all a. For this function, as one can easily check, the normalized intersection of μ_1 and μ_2 is simply μ_2: $\mu_1 \cap \mu_2 = \mu_2$, thus $(\mu_1 \cap \mu_2) \cap \mu_3 = \mu_1 \cap (\mu_2 \cap \mu_3) = \mu_3$.
- The function $t(a, b) = a$ has the property that $t(a, 1) = a$ for all a. For this function, as one can easily check, the normalized intersection of μ_1 and μ_2 is simply μ_1: $\mu_1 \cap \mu_2 = \mu_1$, thus $(\mu_1 \cap \mu_2) \cap \mu_3 = \mu_1 \cap (\mu_2 \cap \mu_3) = \mu_1$.

Proof of the Proposition.

1°. It is known that the normalized intersection operation corresponding to algebraic product is associative. So, to complete the proof, it is sufficient to prove that for each "and"-like operation $t(a, b)$, if the corresponding normalized intersection is associative, then $t(a, b) = a \cdot b$.

2°. Indeed, assume that for an "and"-like operation $t(a, b)$, the corresponding normalized intersection is associative. Let us prove that in this case, $t(a, b) = a \cdot b$.

3°. If $b = 1$, then, by definition of the "and"-like operation, we have $t(a, b) = t(a, 1) = a$ and $a \cdot b = a \cdot 1 = a$, so in this case indeed $t(a, b) = a \cdot b$. Thus, to complete the proof, it is sufficient to consider the case when $b < 1$.

4°. For every two numbers $a, b \in [0, 1]$ for which $b < 1$, let us consider the following three normalized fuzzy sets on the universal set $X = \{1, 2\}$:

$$\mu_1(1) = a, \quad \mu_1(2) = 1, \quad \mu_2(1) = 1, \quad \mu_2(2) = a, \quad \mu_3(1) = 1, \quad \mu_3(2) = b.$$

Let us use associativity to prove that $t(a, b) = a \cdot b$.

4.1°. Let us first compute $(\mu_1 \cap \mu_2) \cap \mu_3$. First, for $\mu_1 \cap \mu_2$, by the definition of the "and"-like operation, we get

$$t(\mu_1(1), \mu_2(1)) = t(a, 1) = a, \quad t(\mu_1(2), \mu_2(2)) = t(1, a) = a,$$

thus

$$\max(t(\mu_1(1), \mu_2(1)), t(\mu_1(2), \mu_2(2)) = \max(a, a) = a.$$

In this case, if $a > 0$, then, by the formula (2), we get

$$(\mu_1 \cap \mu_2)(1) = (\mu_1 \cap \mu_2)(2) = \frac{a}{a} = 1.$$

(The case when $a = 0$ is considered in Part 5 of this proof.) Then, for $(\mu_1 \cap \mu_2) \cap \mu_3$, by the same definition of the "and"-like operator, we get

$$t((\mu_1 \cap \mu_2)(1), \mu_3(1)) = t(1, 1) = 1, \quad t((\mu_1 \cap \mu_2)(2), \mu_3(2)) = t(1, b) = b.$$

Thus,

$$\max(t((\mu_1 \cap \mu_2)(1), t((\mu_1 \cap \mu_2)(2)) = \max(1, b) = 1$$

and therefore, by the formula (3), we have

$$((\mu_1 \cap \mu_2) \cap \mu_3)(1) = \frac{1}{1} = 1, \quad ((\mu_1 \cap \mu_2) \cap \mu_3)(2) = \frac{b}{1} = b. \tag{4}$$

4.2°. Let us now compute $\mu_1 \cap (\mu_2 \cap \mu_3)$. First, from the definition of the "and"-like operation, we get

$$t(\mu_2(1), \mu_3(1)) = t(1, 1) = 1, \quad t(\mu_2(2), \mu_3(2)) = t(a, b).$$

Here, $t(a, b) \le 1$–since all the values of the function t are from the interval [0, 1]. Thus:

$$\max(t(\mu_2(1), \mu_3(1)), t(\mu_2(2), \mu_2(2))) = \max(1, t(a, b)) = 1$$

and therefore, by the formula (2),

$$(\mu_2 \cap \mu_3)(1) = \frac{1}{1} = 1, \quad (\mu_2 \cap \mu_3)(2) = \frac{t(a, b)}{1} = t(a, b).$$

Now, from the definition of the "and"-like operation, we get

$$t(\mu_1(1), t(\mu_2(1), \mu_3(1))) = t(a, 1) = a,$$

$$t(\mu_1(2), t(\mu_2(2), \mu_3(2))) = t(1, t(a, b)) = t(a, b). \tag{5}$$

According to the formula (3), to find the values of the fuzzy set $\mu_1 \cap (\mu_2 \cap \mu_3)$, we need to find the maximum of the two values (5). The value of this maximum depends on which of the two values (5) is larger. Let us consider both possible cases.

4.2.1°. If $a \le t(a, b)$, then

$$\max(t(\mu_1(1), t(\mu_2(1), \mu_3(1))), t(\mu_1(2), t(\mu_2(2), \mu_3(2))) = \max(a, t(a, b)) = t(a, b)$$

and thus, by the formula (3), we get

$$(\mu_1 \cap (\mu_2 \cap \mu_3))(1) = \frac{a}{t(a,b)}, \quad (\mu_1 \cap (\mu_2 \cap \mu_3))(2) = \frac{t(a,b)}{t(a,b)} = 1.$$

By associativity, these values should be equal to the values (4). By comparing the values of these two fuzzy sets for $x = 2$, we conclude that $b = 1$ which contradicts to our assumption that $b < 1$. Thus, this case is impossible.

4.2.2°. Since, as we have proven, we cannot have $a \le t(a,b)$, we must have $t(a,b) < a$. In this case,

$$\max(t(\mu_1(1), t(\mu_2(1), \mu_3(1))), t(\mu_1(2), t(\mu_2(2), \mu_3(2)))) = \max(a, t(a,b)) = a$$

and thus, by the formula (3), we get

$$(\mu_1 \cap (\mu_2 \cap \mu_3))(1) = \frac{a}{a} = 1, \quad (\mu_1 \cap (\mu_2 \cap \mu_3))(2) = \frac{t(a,b)}{a}.$$

By associativity, these values should be equal to the values (4). By comparing the values of these two fuzzy sets for $x = 2$, we conclude that

$$\frac{t(a,b)}{a} = b,$$

hence $t(a,b) = a \cdot b$.

For the values $a > 0$, the proposition is proven.

5°. Let us now consider the case when $a = 0$. In this case, by the formulas from Part 4.1 of this proof, the normalized intersection $\mu_1 \cap \mu_2$ is undefined, and, thus, the set $(\mu_1 \cap \mu_2) \cap \mu_3$ is undefined as well. Thus, by associativity, the set $\mu_1 \cap (\mu_2 \cap \mu_3)$ should also be undefined. According to the formulas from Part 4.2 of this proof, we get

$$(\mu_2 \cap \mu_3)(1) = 1, \quad (\mu_2 \cap \mu_3)(2) = t(0,b).$$

Now, from the definition of the "and"-like operation, we get

$$t(\mu_1(1), t(\mu_2(1), \mu_3(1))) = t(0,1),$$

$$t(\mu_1(2), t(\mu_2(2), \mu_3(2))) = t(1, t(0,b)) = t(0,b). \tag{6}$$

The fact that the fuzzy set $\mu_1 \cap (\mu_2 \cap \mu_3)$ is undefined means that the corresponding denominator is equal to 0:

$$\max(t(\mu_1(1), t(\mu_2(1), \mu_3(1))), t(\mu_1(2), t(\mu_2(2), \mu_3(2))) = \max(t(0,1), t(0,b)) = 0.$$

Thus,

$$0 \le t(0,b) \le \max(t(0,1), t(0,b)) = 0,$$

so indeed $t(0, b) = 0 \cdot b$.

So, the proposition is also proven for the remaining case $a = 0$.

Comment. In our proof, the only assumption that we made about the operation $t(a, b)$ is that $t(a, 1) = t(1, a) = a$ for all a. If we additionally assumed that $t(a, b)$ is a t-norm, then the proof would be much shorter; indeed:

- for a t-norm we always have $t(a, b) \le a$, so there is no need to consider the case $t(a, b) > a$, and
- we always have $t(0, b) = 0$, so there is no need to prove this formula.

Acknowledgements This work was supported in part by the National Science Foundation grants 1623190 (A Model of Change for Preparing a New Generation for Professional Practice in Computer Science), HRD-1834620 and HRD-2034030 (CAHSI Includes), EAR-2225395 (Center for Collective Impact in Earthquake Science C-CIES), and by the AT&T Fellowship in Information Technology. It was also supported by the program of the development of the Scientific-Educational Mathematical Center of Volga Federal District No. 075-02-2020-1478, and by a grant from the Hungarian National Research, Development and Innovation Office (NRDI). The authors are thankful to all the participants of the 2023 International Conference on Fuzzy Systems FUZZ-IEEE 2023 (Incheon, Korea, August 13–17, 2023) for valuable discussions.

References

1. Belohlavek, R., Dauben, J.W., Klir, G.J.: Fuzzy Logic and Mathematics: A Historical Perspective. Oxford University Press, New York (2017)
2. Dubois, D., Prade, H., Yager, R.: Merging fuzzy information. In: Bezdek, J.C., Dubois, D., Prade, H. (eds.) Fuzzy Sets in Approximate Reasoning and Information Systems, pp. 335–401. Kluwer Academic Publishers, Boston (1999)
3. Klir, G., Yuan, B.: Fuzzy Sets and Fuzzy Logic. Prentice Hall, Upper Saddle River, New Jersey (1995)
4. Mendel, J.M.: Uncertain Rule-Based Fuzzy Systems: Introduction and New Directions. Springer, Cham, Switzerland (2017)
5. Nguyen, H.T., Walker, C.L., Walker, E.A.: A First Course in Fuzzy Logic. Chapman and Hall/CRC, Boca Raton, Florida (2019)
6. Novák, V., Perfilieva, I., Močkoř, J.: Mathematical Principles of Fuzzy Logic. Kluwer, Boston, Dordrecht (1999)
7. Zadeh, L.A.: Fuzzy sets. Inf. Control **8**, 338–353 (1965)

High Potential Negative Sampling for Drug Disease Association Prediction

Manh Hung Le, Nam Anh Dao, and Xuan Tho Dang

Abstract As we are aware, drug development incurs substantial costs and time investments. Thus, there is a need for novel methods to efficiently support this process, such as drug repositioning. This field is also gaining significant attention from researchers, particularly with the aid of intelligent computational methods applied in machine learning. These methods open new opportunities in this domain and have yielded promising outcomes. However, a major drawback of learning lies in the necessity for constructing a well-supported dataset of positive and negative samples for prediction. Current methods largely focus on generating randomly sampled negative instances. Clearly, such randomly sampled negative instances cannot ensure the quality of negative dataset, and they undoubtedly impact the prediction performance of classifiers. The article proposes a method of selecting highly reliable negative samples for performance improvement in prediction of drug-disease association. Experimental results demonstrate that the proposed method outperforms current research, providing an appealing alternative solution for predicting drug-disease association.

Keywords Drug disease association · High potential negative · Meta-path · Protein protein association · Drug repositioning

1 Introduction

Exploring and developing new drugs is a relatively complex process that demands substantial financial investment and time, coupled with a significant risk of failure. In this context, this article focuses on a potential strategy that could reshape this landscape–drug repurposing. Reports suggest that introducing a new small molecule drug to the market can cost between \$0.31 to \$2.8 billion and take around 10 to 17

M. Hung Le · N. A. Dao
Electric Power University, Hanoi, Vietnam
e-mail: anhdn@epu.edu.vn

X. T. Dang (✉)
Academy of Policy and Development, Hanoi, Vietnam
e-mail: thodx@apd.edu.vn

N. Hoang Phuong et al. (eds.), *Machine Learning and Other Soft Computing Techniques: Biomedical and Related Applications*, Studies in Systems, Decision and Control 543, https://doi.org/10.1007/978-3-031-63929-6_7

years [1]. The failure rate in Phase I clinical trials can reach nearly 90% [2]. Given the challenges faced in drug development, the pursuit of ways to enhance efficiency and optimize this process has become paramount.

Against this backdrop, the interest in drug re-purposing is burgeoning. Drug re-purposing proposes the reuse of previously approved drugs for new indications. This not only helps reduce time and resources but also holds the potential to enhance the success rate of drug development. A pivotal factor in drug re-purposing lies in the amalgamation and utilization of existing data. Currently, around 2,589 small molecule drugs have been registered in DrugBank [3]. Simultaneously, in the realm of healthcare, the UMLS database [4] records over 25,000 diseases, generating more than 60 million drug-disease pairs. However, only less than 5% of these have been identified to have a clear therapeutic relationship [5]. This presents an intriguing and promising dimension for applying data analysis methods for uncovering potential associations between drugs and diseases.

With the advancement of intelligent computational methods, there is an approach utilizing machine learning algorithms to construct models capable of predicting relationships between drugs, genes, and diseases. These models are trained on datasets generated by experimental methods, and subsequently utilized to predict associations among drugs, genes, and diseases. The assessment of predictive models' performance is carried out from two perspectives: the accuracy of the models and the validation of prediction outcomes against the knowledge published in medical or biological databases. Though the predictive outcomes may not be entirely accurate, they offer an alternative way that aids bio-informatics researchers in narrowing down subjects for investigation, thus saving time and costs in exploring the roles of drugs, genes, and diseases.

In recent times, the research community has been focusing efforts on the development of computational models to unearth latent knowledge embedded within biological data. Of particular significance is the issue of predicting relationships between different drugs and diseases, which has garnered substantial attention within the field of computer science. Two primary approaches are employed in constructing predictive models. The first approach relies on the modeling of biological networks, while the second approach capitalizes on the application of machine learning techniques. Both approaches are grounded on the hypothesis that if the interaction between a drug and a disease is intricately connected to other components within the biological network, then the plausibility of a relationship between them exists [6, 7]. Among these two aforementioned approaches, the machine learning-based method brings forth notable advantages. Most notably, it possesses the capability to optimize predictive potential by leveraging advanced machine learning algorithms. Recently, numerous predictive models have been proposed, harnessing the potential of machine learning to unveil concealed associations between various drugs and diseases.

Nevertheless, the efficacy of machine learning methods heavily relies on the quality of the training dataset, typically comprising positive and negative samples. In this context, positive samples pose no specific issue. The challenge lies in the scarcity of experimentally verified negative samples. Recent endeavors [8, 9] have addressed this aspect by randomly pairing drugs and diseases and subsequently excluding these pairs from the set of positive samples. In this document, we identify this method as a random negative sampling approach or a traditional approach. Clearly, randomly generated negative samples cannot be considered authentic negative instances. Consequently, the quality of randomly sampled negative datasets cannot be ensured. This undoubtedly impacts the predictive performance of classifiers trained on datasets containing such randomly sampled negative instances.

This article addresses the approach of selecting highly reliable negative samples to enhance the prediction of relationships among drugs, genes, and diseases. The primary contributions of this research can be summarized as follows: (a) We propose a more effective method for sampling negative instances. By carefully controlling these negative samples, we ensure strong and reliable classification performance. (b) Our proposed model achieves superior predictive performance in drug-disease relationship prediction. This achievement marks a significant advancement in this field. (c) Our method demonstrates effectiveness and exceptional accuracy when compared to existing methods. Through rigorous evaluation and comparative analysis, we substantiate the superiority of our proposed approach, providing an appealing alternative solution for drug-disease relationship prediction tasks. (d) The deployment of a more reliable drug-disease classification system brings several benefits, including reducing the clinical workload for physicians, cost savings in drug-disease profiling, and providing accurate diagnoses supported by computational assistance.

The remaining sections of this article are structured as follows: In Sect. 2, we provide an evaluation of existing literature relevant to this topic, examining the strengths and weaknesses of previous methods. Section 3 outlines our proposed methodology, delineating the key components and strategies employed. Experimental results are presented and analyzed in Sect. 4, affording an in-depth insight into the performance and efficacy of our approach. Finally, in Sect. 5, we summarize our findings, draw conclusions based on the results, and discuss potential directions for future research and development.

2 Related Work

Through our research in the field of predicting drug-disease relationships and drug repositioning, we have identified several distinct approaches. These strategies involve harnessing information derived from studies and databases to unveil potential associations between drugs and diseases. For instance, a noteworthy investigation conducted by Sirota et al. [12] analyzed gene expression data from 100 diseases and 164 drugs, uncovering promising indications for these medications. One intriguing discovery was the prospective efficacy of cimetidine, a commonly prescribed drug for peptic ulcers, in treating lung adenocarcinoma [11]. This exemplifies the utility of gene expression profiles in identifying drugs with potential applications across various diseases.

It is noteworthy that the utilization of gene-disease relationships in current research has garnered substantial attention from numerous investigators, encompassing machine learning [13–16], and deep learning methods [8, 9, 17, 18]. These methodologies show promise in enhancing our comprehension of drug-disease associations and fostering progress in the field. These methodologies show promise in enhancing our comprehension of drug-disease associations and fostering progress in the field.

Machine Learning Methods. Moghadam et al. [13] introduced the cored Mean Kernel Fusion(SMKFS) technique for forecasting drug-disease associations within the domain of drug repositioning. This approach harnesses diverse data sources and features to enhance prediction accuracy and reliability. SMKF merges drug and disease features at both the drug-drug and drug-disease levels via a scored mean kernel fusion approach.

Liang et al. [14] proposed the Laplacian Regularized Sparse Subspace Learning (LRSSL) method to predict drug indications and extract significant drug features. This methodology combines the projection of drug features into a shared subspace, employs Laplacian regularization for constraint, and integrates sparse feature selection to enhance model performance. Huang et al. [15] put forth a method leveraging drug structure information and target protein sequences to construct a predictive model. The sample dataset was segregated into known drug-target interaction pairs and randomly chosen pairs. Drugs are represented using molecular substructure fingerprints, while protein sequences are encoded via the Pseudo-SMR technique. The prediction model is constructed utilizing the Randomized Trees algorithm for various protein target types.

Deep Learning Methods. Liu et al. [8] introduces a heterogeneous network-based deep neural network model termed HNet-DNN for predicting novel drug-disease associations. The model employs raw features of drugs and diseases to construct drug-drug and disease-disease similarity networks, which are subsequently fused with established drug-disease associations to constitute a heterogeneous network. Topological features derived from this network serve as input vectors for the deep neural network, resulting in enhanced prediction performance compared to conventional methods.

Wang et al. [9] presents HNRD, a neural-network-based approach for forecasting drug-disease associations within a heterogeneous network. The technique generates a heterogeneous network by amalgamating neighborhood information and then learns node representations of drugs and diseases to match observed matrices. HNRD attains high performance in cross-validation experiments, outperforms state-of-the-art methods, and demonstrates promise in leave-one-out cross-validation experiments. Jiang et al. [17] proposes a computational model that combines a sparse auto-encoder with a rotation forest to predict drug-disease associations. This model utilizes a fusion technique to acquire combined features from multiple sources and incorporates a feature extraction module grounded in a sparse autoencoder and Principal Component Analysis (PCA). Subsequently, the extracted features are fed into the rotation forest ensemble classifier for final prediction, where higher prediction scores indicate a higher likelihood of drug-disease associations.

Most existing methods for predicting drug-disease relationships utilize labeled drugs as positive samples and randomly select unlabeled drugs as negative samples. However, this approach may inadvertently incorporate true positive samples within the unlabeled drugs, leading to inaccurate selection of negative samples and diminished prediction performance. To address this concern, Dang et al. [6] and Wu et al. [7] introduced the integration of various interaction data types and the consideration of diverse dependencies using metapaths. They reduced the dimensionality of each meta-path through the Singular Value Decomposition (SVD) algorithm and trained the model utilizing the Random Forest algorithm. These studies employed a filter to discern negative samples based on the premise that if a drug and disease are not associated with the same protein, they are unrelated to each other.

Although the above methods have improved the selection of negative samples compared to random selection, they still fail to encompass all potential latent relationships between drugs and diseases. Therefore, in this study, we propose a novel method called the high-potential negative sampling to enhance the quality of negative samples. This method considers the relationships between protein pairs, in addition to the interactions between drugs and proteins and disease-protein interactions. By incorporating these y relationships, optimized negative samples are chosen, leading to improved predictive efficacy.

3 The Method

Data for the drug-disease problem covers disease-protein interactions, drug-protein interactions, and disease-drug interactions. We collected known drug-disease pairs as positive samples and created three datasets using three methods: random selection, filtering, and our improved filtering method. These datasets were then used to evaluate the performance of predicting drug-disease relationships using classifiers.

Figure 1 illustrates the workflow of our method, which comprises five steps: (1) Constructing a heterogeneous drug-protein-disease network based on previously known relationships of drug-protein, disease-protein, and drug-disease interactions; (2) Generating high-potential negative datasets using random selection, filtering, and our improved method; (3) Extracting five meta-path patterns from the heterogeneous network to support feature vector construction for each drug-disease pair in step; (4) Training classifiers using an Ensemble classifier; (5) Evaluating performance.

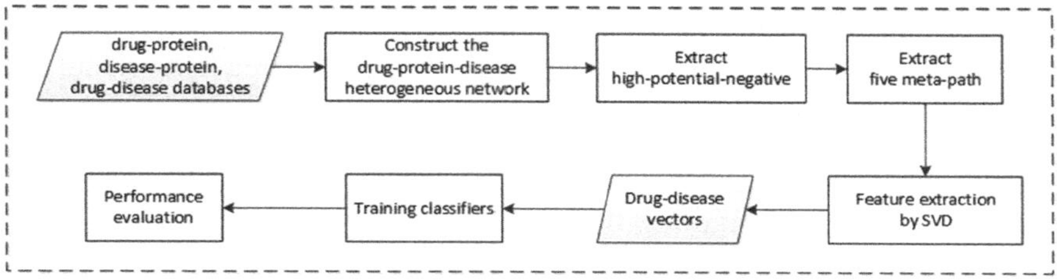

Fig. 1 The workflow consists of five fundamental steps. Rectangles depict the executive modules, while parallelograms represent the data modules

Construct the drug-protein-disease heterogeneous network. The work uses the notations d for drugs, s for diseases, and p for proteins. We represent the interactions between drugs and diseases using matrix $A_{ds}[m \times n]$. Each element $A_{ds}[i, j]$ is set to 1 if drug d_i treats disease s_j and $A_{ds}[i, j] = 0$ if the relationship between drug d_i and disease s_j is unknown. Similarly, the dataset for drug-protein interactions is denoted as $A_{dp}[n \times z]$, where $A_{dp}[i, j] = 1$ if drug d_i contains protein p_j, and vice versa, $A_{dp}[i, j] = 0$ if there is no such interaction. The disease-protein dataset is represented as $A_{sp}[n \times z]$, where $A_{sp}[i, j] = 1$ if disease s_i and protein p_j have a direct relationship, and $A_{sp}[i, j] = 0$ if there is no such relationship.

Extract high potential negative samples. When using Bayesian inference [10] to represent the drug disease interaction for drug design, it is commonly appreciated that prediction $P(d|s)$ of using a drug d for a disease s is evaluated, particularly by likelihood $P(s|d)$ of p given s, prior probability $P(d)$ and marginal probability $P(s)$:

$$P(d|s) = P(s|d)P(d)/P(s). \tag{1}$$

However, the marginal probability $P(s)$ of the disease s is usually can be ignored leaving question for the likelihood $P(s|d)$ and the prior probability $P(d)$:

$$P(d|s) = P(s|d)P(d). \tag{2}$$

In particular the drug disease interaction, which has data in the past to be analyzed, would be beneficial for estimating the prior probability $P(d)$ from available training data:

$$P(d) = \sum_{s} P(d|s)P(s). \tag{3}$$

When constructing the drug-protein-disease heterogeneous network, different values of $P(d|s)$ from (3) can be used to find the prior probability $P(d)$ with value of zero or one:

$$P(d|s) = \begin{cases} 1 \\ 0 \end{cases} \tag{4}$$

It is also of interest to note that number of drug-disease interactions which were confirmed to be fine coded by 1 is very small in comparison with the number of non-confirmed drug-disease interactions marked by 0 in (4). In the training task, samples are selected in the way that the number of positive samples with value of 1 is equal the number of negative samples with value of 0. This is required for avoiding poor predictive performance:

$$count_{P(d|s)=1}(ds\ pair) = count_{P(d|s)=0}(ds\ pair). \tag{5}$$

Whereas our heterogeneous network provides also interactions of drug–protein by $P(d|p)$, protein–protein by $P(p|p)$ and protein–disease by $P(p|s)$, the drug–disease interaction $P(d|s)$ can have other way of looking for estimation:

$$P(d|s) = P(d|p)P(p|p)P(p|s). \tag{6}$$

From this viewpoint, we find that Eq. (6) can result in basic values of 0, 1 and other values (k) in interval (0, 1):

$$P(d|s) = \begin{cases} 1 \\ 0 \\ k \in (0, 1) \end{cases} \tag{7}$$

Since we are interested in selecting negative samples from a large number of negative samples, it can be suitable to choose the negative samples that have $P(d|s)$ estimated exactly by zeros removing all samples with $k \in (0, 1)$. The selection of the negative samples can help to improve correctness of $P(d)$ estimation in Eq. (3) and so enhance the prediction performance. So, our method for constructing negative samples does not directly utilize the drug-disease interaction but uses disease-protein interactions and drug-protein interactions.

The concept can be also explained by observations: if a drug d_i and a disease s_j share a common protein p^t, then the drug-disease pair (d_i, s_j) is likely to be positive. Conversely, if a drug-disease pair (d_i, s_j) does not share direct protein overlap, it is likely to be negative. In the case where a drug d_i and a disease s_j are potentially negative but are still connected through protein interactions, the negativity likelihood is not high. On the contrary, if a drug d_i and a disease s_j are not linked through any pathways, the likelihood of being negative is substantial.

Figure 2 depicts the relationship between drugs and diseases mediated by proteins. The drug-disease pair (d_1, s_1) demonstrates a positive likelihood, whereas the pair (d_1, s_2) exhibits a negative likelihood. The pairs (d_1, s_2) and (d_2,s_1) carry a notably high likelihood of negativity.

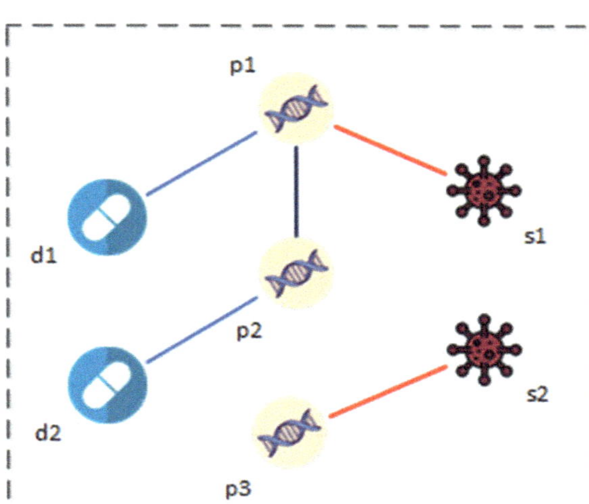

Fig. 2 The heterogeneous biological network describes the relationships among drug-protein-disease entities

Motivated by the mentioned above concept, we select high-potential-negative samples for training as follows. Firstly, we propose to construct a protein-protein interaction matrix A_{pp} to represent the relationships among proteins, based on the drug-protein matrix and disease-protein matrix. To have the matrix, we establish the protein-protein association matrix A_{ppd}, which records the connections between proteins using the drug-protein matrix. Likewise, we construct the A_{pps} matrix, reflecting protein relationships based on the disease-protein matrix. Subsequently, the A_{pp} matrix is formed through the linear combination of A_{ppd} and A_{pps} matrices. The formulas of the A_{ppd}, A_{pps} and A_{pp} are defined in (8), (9) and (10), respectively.

$$A_{ppd} = A_{dp}^{T} A_{dp} \tag{8}$$

$$A_{pps} = A_{sp}^{T} A_{sp} \tag{9}$$

$$A_{pp} = A_{ppd} A_{pps} \tag{10}$$

These calculations enable us to construct the A_{pp} protein-protein matrix, representing protein relationships based on the drug-protein and disease-protein matrices.

Next, the correlation between drugs and diseases is re-evaluated by employing the linear multiplication of three matrices: the drug-protein matrix, the disease-protein matrix, and the protein-protein matrix. The process of extracting high-potential negatives is outlined in Algorithm 1.

Algorithm 1: High potential negative samples

```
Input   : A_dp[d × p] (drug-protein-matrix)
          A_sp[s × p] (disease-protein-matrix)
          A_pp[p × p] (protein-protein-matrix)
Output  : H (Negative samples set )
Begin Algorithm
A_ds ← A_dp A_pp A_sp^T, See equation (6)
H ← Ø;
while i ≤ m do
    while j ≤ n do
        if A(i, j) = 0 then
            H ← H ∪ {(i, j)};
        end
        j ← j + 1;
    end
    i ← i + 1;
end
return H;
End Algorithm
```

Extracting five meta-paths and performing feature extraction using SVD. Within a heterogeneous network, a meta-path denotes a sequence of node and link types designed to encapsulate the collective relationships existing between two distinct node types. It encompasses both the structural attributes and semantic associations inherent in the network, thereby facilitating the extraction of pertinent pathways connecting pairs of nodes. Meta-paths offer a flexible and robust framework for comprehending and structurally modeling intricate interactions within heterogeneous networks.

In this research, we employ five distinct meta-paths, as introduced by Wu et al. [7], labeled as MP1, MP2, MP3, MP4, and MP5 with formulas (11)–(15):

Meta-Path 1: Drug →Disease

$$MP1 = A_{ds}, \tag{11}$$

Meta-Path 2: Drug → Protein → Disease

$$MP2 = A_{dp}A_{sp}^T, \tag{12}$$

Meta-Path 3: Drug → Protein → Drug → Disease

$$MP3 = A_{dp}A_{dp}^T A\_ds, \tag{13}$$

Meta-Path 4: Drug → Disease → Drug → Disease

$$MP4 = A_{ds}A_{ds}^T A_{ds}, \tag{14}$$

Meta-Path 5: Drug → Disease → Protein → Disease

$$MP5 = A_{ds}^T A_{sp}A_{sp}^T. \tag{15}$$

These meta-paths provide diverse perspectives to capture different aspects of the interactions between drugs, proteins, and diseases. By leveraging the structural and semantic information encoded in these meta-data, we can develop efficient models to analyze and predict the relationships between diseases and drugs within heterogeneous networks.

We have obtained five drug-disease interaction matrices, namely MP1, MP2, MP3, MP4, and MP5, through five meta-paths. Each matrix has dimensions of (1186 × 449), representing varying degrees of similarity between drugs and diseases. These matrices capture different strengths of pathways. However, the size of these matrices is large, and the data sparsity is high. To address this issue, we applied Singular Value Decomposition (SVD) to reduce the dimensionality. By selecting the top eigenvalues, we retained the most crucial information in the matrix while eliminating redundant information. This allowed us to construct an efficient prediction model for drug-disease interactions with reduced feature dimensions. Moreover, the matrix size was reduced through this process.

Next, we employed a supervised machine learning model based on the Random Forest (RF) classifier to predict drug-disease interactions for each dataset corresponding to each meta-path. Finally, an ensemble classifier was utilized to achieve strong performance across the five RF models built in the previous step.

Performance Evaluation. In data analysis, it is significant to evaluate the performance of prediction models. To evaluate the performance of the new dataset, we divide the dataset into 5 parts and perform cross-validation. Accuracy (ACC) in (16), Recall (REC) in (17), Precision (PRE) in (18), Matthews Correlation Coefficient (MCC) in (19), F1-score in (20) and G-mean in (21) is the most important metrics used for performance evaluation.

$$ACC = \frac{TP + TN}{TP + FP + TN + FN} \tag{16}$$

$$REC = \frac{TP}{TP + FN} \tag{17}$$

$$PRE = \frac{TP}{TP + FP} \tag{18}$$

$$MCC = \frac{TP \times TN - FP \times FN}{\sqrt{(TP + FP)(TP + FN)(TN + FP)(TN + FN)}} \tag{19}$$

$$F1 - score = \frac{2 \times PRE \times REC}{PRE + REC} \tag{20}$$

$$G - mean = \sqrt{\text{SP} \times \text{SE}} \tag{21}$$

where TP, TN, FP, FN, SP and SE denote the number of true positive samples, true negative samples, false positive samples, false negative samples, specificity and sensitivity respectively. Sensitivity and Specificity are two crucial terms in the performance evaluation of classification models, with $SP = TP/(TP + FN)$ and $SP = TN/(FP + TN)$.

4 Experiments

In this study, we extracted data from three independent sources: disease-protein interactions from OMIM [19], drug-protein interactions from Gottlieb [20], and disease-drug interactions from DrugBank [3]. The number of drugs is denoted as m, with $m = 1186$ drugs, the number of diseases as n, with $n = 449$ diseases, and the number of proteins as z, with $z = 1467$ proteins.

Table 1 presents the quantitative descriptions of the three datasets. Specifically, there are 1827 approved drug-disease pairs and 657,318 unapproved pairs. The

Table 1 Description of experimental dataset

Dataset source	Relation	Matrix size	Number of interactions	Number of interactions	Ratio in %
OMIN	Disease-protein	(449, 1,467)	1,365	657,318	0.207
DrugBank	Drug-protein	(1,186, 1,467)	4,642	1,735,220	0.267
Gottlieb	Drug-disease	(1,186, 449)	1,827	530,687	0.344

Table 2 Comparison of negative samples construction methods

Methods	AUPR	AUC	PRE	REC	ACC	MCC	F1	G-mean
Random	0.954	0.949	0.910	0.848	0.872	0.747	0.877	0.854
Filter	0.956	0.950	0.912	0.855	0.875	0.753	0.882	0.855
Our method	**0.968**	**0.965**	**0.930**	0.882	**0.902**	**0.806**	**0.905**	**0.868**

*The best scores are printed in bold

dataset for drug-protein interactions includes 4624 interacting pairs and 1,735,220 non-interacting pairs. For protein-disease interactions, there are 1365 confirmed pairs and 530,687 unconfirmed pairs. From Table 1, it is evident that there are 1827 positive samples, which correspond to known drug-disease pairs. Regarding most biomedical data, negative samples are drawn from a pool of 657,318 undetermined disease pairs. However, utilizing these negative samples poses a significant challenge since the 657,318 pairs are experimentally unverified, suggesting the possibility of missing potential drug-disease relationships. In the random negative sampling approach, negative samples are randomly selected from the pool of 657,318 unknown pairs, yielding unreliable results. While filtering methods [6, 7] improve negative sample selection, we contend that this method is relatively straightforward.

Table 2 shows the results of three negative samples extraction methods, where the first six rows are the results of using the random negative method. Here the first five lines represent the results of meta-path-1 to meta-path-5 respectively, the sixth line is the ensemble result of the model. The subsequent six lines showcase the outcomes of the samples set extracted using the negative filter method. Finally, six lines portray the performance of our method. Our new dataset yields superior results in six out of seven metrics compared to the other methods: AUPR = 96.81%, AUC = 96.58%, PRE = 93.02%, REC = 88.29%, ACC = 90.26%, MCC = 80.63%, and F1 = 90.57% Geometric G-mean = 0.868%. The results are illustrated in Fig. 3. This ensured that our sampling method is reliable and capable for accurate prediction.

For each dataset, we conducted a 5-fold cross-validation, wherein the data was divided into five subsets for evaluation. This cross-validation was carried out with a 1:4 ratio, signifying that 20% of the data was designated as the test set, while the remaining 80% constituted the training set. In Fig. 4a, the AUC metric for the negative random dataset is depicted. This figure exhibits five distinctive curves, with each curve representing the AUC resulted from individual folds marked by unique color. Similarly, Fig. 4b shows the AUPR measure for the negative random selection method, featuring five curves that represent the AUPR outcomes for each fold with

Table 3 Performance of related methods

Method	Database	AUPR	AUC	PRE	REC	ACC	MCC	F1
LRSSL [14]	PubChem [35], InterPro [36], UniProt [37], SIDER [38]	0.881	0.861	0.864	0.732	0.770	0.553	0.790
PREDICT [25]	DrugBank [3], repoDB [34]	0.908	0.895	0.809	0.850	0.830	0.662	0.828
MBiRW [21]	OMIM [32], DrugBank [33]	0.952	0.942	0.867	0.901	0.884	0.769	0.884
EMP-SVD [24]	OMIM [19], Gottlieb [20], DrugBank [3]	0.956	0.951	0.913	0.854	0.876	0.755	0.882
New Metapatht [6]	OMIM [19], Gottlieb [20], DrugBank [3]	0.962	0.956	0.882	**0.901**	0.889	0.780	0.889
Our method	OMIM [19], Gottlieb [20], DrugBank [3]	**0.968**	**0.966**	**0.930**	0.882	**0.903**	**0.806**	**0.906**

*The best scores are printed in bold

potential for treatment for the disease. This finding was documented in a study by [24] and confirmed in subsequent studies conducted by [26–28].

Regarding the disease of Non-small somatic cell lung cancer we discovered two drugs, Carboplatin and Cisplatin, that are effective in its treatment. Initially, used in the treatment of advanced ovarian carcinoma [39]. Cisplatin, on the other hand, was originally prescribed for malignant tumors of the urinary bladder, esophagus, and others. In 1992, Saito et al. conducted a clinical trial involving 17 patients in Stage 1, assessing the clinical efficacy of Carboplatin and Cisplatin [29] and conducted a Phase II clinical trial in 1993. In 1998, J.-P. Sculier et al. conducted a phase III trial involving 529 patients [30], which yielded promising results [31]. Therefore, the use of Carboplatin and Cisplatin for the treatment of non-small cell lung cancer is considered reliable.

5 Conclusions

In this paper, we introduced a method called High Potential Negative Sampling (HPNS) for selecting reliable negative samples based on the relationships between drug-protein-protein-disease. Our method demonstrates effectiveness and accuracy, particularly when compared to existing methods. Through rigorous evaluation and comparative analysis, we showcased the superiority of our proposed method, offering an appealing alternative for predicting drug-disease relationships. The findings from the case study reaffirm the reliability and utility of our approach in identifying associations between drugs and diseases, providing valuable insights for future research and promising therapeutic applications.

Acknowledgements The paper is the result of the scientific research project at the Electric Power University in 2023.

References

1. Rennane, S., Baker, L., Mulcahy, A.: Estimating the cost of industry investment in drug research and development: a review of methods and results. Inquiry **58**, 469580211059731 (2021)
2. Wu, G., Liu, J., Wang, C.: Predicting drug-disease interactions by semi-supervised graph cut algorithm and three-layer data integration. BMC Med. Genomics **10**(Suppl 5), 79 (2017)
3. Wishart, D.S., Feunang, Y.D., Guo, A.C., Lo, E.J., Marcu, A., Grant, J.R., Sajed, T., Johnson, D., Li, C., Sayeeda, Z., Assempour, N., Iynkkaran, I., Liu, Y., Maciejewski, A., Gale, N., Wilson, A., Chin, L., Cummings, R., Le, D., Pon, A., Knox, C., Wilson, M.: DrugBank 5.0: a major update to the DrugBank database for 2018. Nucl. Acids Res. 46 (D1): D1074–D1082 (2018)
4. Gabetta, M., Larizza, C., Bellazzi, R.: A Unified Medical Language System (UMLS) based system for Literature-Based Discovery in medicine. Stud. Health Technol. Inform. **192**, 412–416 (2013). PMID: 23920587
5. Hurle, M.R., Yang, L., Xie, Q., Rajpal, D.K., Sanseau, P., Agarwal, P.: Computational drug repositioning: from data to therapeutics. Clin. Pharmacol. Ther. **93**(4), 335–41 (2013). PMID: 23443757
6. Dang, X.T., Le, M.H., Dao, N.A.: Drug repositioning for drug disease association in Meta-paths. In: Deep Learning and Other Soft Computing Techniques. Studies in Computational Intelligence, vol. 1097. Springer (2023)
7. Wu, G., Liu, J., Yue, X.: Prediction of drug-disease associations based on ensemble meta paths and singular value decomposition. BMC Bioinform. **20**(Suppl 3), 134 (2019)
8. Liu, H., Zhang, W., Song, Y., Deng, L., Zhou, S.: HNet-DNN: inferring new drug-disease associations with deep neural network based on heterogeneous network features. J. Chem. Inf. Model. **60**(4), 2367–2376 (2020)
9. Wang, Y., Deng, G., Zeng, N., Song, X., Zhuang, Y.: Drug-disease association prediction based on neighborhood information aggregation in neural networks. IEEE Access **7**, 50581–50587 (2019)
10. Barber, D.: Bayesian Reasoning and Machine Learning. Cambridge University Press (2012)
11. Li, Y., Qiu, C., Tu, J., Geng, B., Yang, J., Jiang, T., Cui, Q.: HMDD v2.0: a database for experimentally supported human microRNA and disease associations. Nucl. Acids Res. **42**(Database issue), D1070–D1074 (2014)
12. Sirota, M., Dudley, J.T., Kim, J., Chiang, A.P., Morgan, A.A., Sweet-Cordero, A., Sage, J., Butte, A.J.: Discovery and preclinical validation of drug indications using compendia of public gene expression data. Sci. Transl. Med. **3**(96), 96ra77, 17 Aug 2011. https://doi.org/10.1126/scitranslmed.3001318. Erratum in: Sci. Transl. Med. **3**(102), 102er7, 28 Sep 2011 (2001)
13. Moghadam, H., Rahgozar, M., Gharaghani, S.: Scoring multiple features to predict drug disease associations using information fusion and aggregation. SAR QSAR Environ. Res. **27**(8), 609–28 (2016)
14. Liang, X., Zhang, P., Yan, L., Fu, Y., Peng, F., Qu, L., Shao, M., Chen, Y., Chen, Z.: LRSSL: predict and interpret drug-disease associations based on data integration using sparse subspace learning. Bioinformatics **33**(8), 1187–1196 (2017)
15. Huang, Y.A., You, Z.H., Chen, X.: A systematic prediction of drug-target interactions using molecular fingerprints and protein sequences. Curr. Protein Pept. Sci. **19**(5), 468–478 (2018)
16. Bolgár, B., Antal, P.: VB-MK-LMF: fusion of drugs, targets and interactions using variational Bayesian multiple kernel logistic matrix factorization. BMC Bioinform. **18**, 440 (2017)

17. Jiang, H.J., Huang, Y.A., You, Z.H.: SAEROF: an ensemble approach for large-scale drug-disease association prediction by incorporating rotation forest and sparse autoencoder deep neural network. Sci. Rep. **10**, 4972 (2020)
18. Jiang, H.J., Huang, Y.A., You, Z.H.: Predicting drug-disease associations via using gaussian interaction profile and kernel-based autoencoder. BioMed Res. Int. **2019**, 2426958 (2019)
19. Hamosh, A., Scott, A.F., Amberger, J.S., Bocchini, C.A., McKusick, V.A.: Online Mendelian Inheritance in Man (OMIM), a knowledgebase of human genes and genetic disorders. Nucl. Acids Res. **33**(Database issue), D514–D517 (2005)
20. Gottlieb, A., Stein, G.Y., Ruppin, E., Sharan, R.: PREDICT: a method for inferring novel drug indications with application to personalized medicine. Mol. Syst. Biol. **7**, 496 (2011)
21. Luo, H., et al.: Drug repositioning based on comprehensive similarity measures and Bi-Random walk algorithm. Bioinformatics **32**(17), 2664–2671 (2016)
22. O'Boyle, N.M., Banck, M., James, C.A., et al.: Open Babel: An open chemical toolbox. J. Cheminform. **3**, 33 (2011)
23. Smith, T.F., Waterman, M.S., Burks, C.: The statistical distribution of nucleic acid similarities. Nucl. Acids Res. **13**(2), 645–56 (1985)
24. Wu, G., Liu, J., Yue, X.: Prediction of drug-disease associations based on ensemble meta paths and singular value decomposition. BMC Bioinform. **20**(Suppl 3), 134 (2019)
25. Zeng, X., Zhu, S., Liu, X., et al.: deepDR: a network-based deep learning approach to in silico drug repositioning. Bioinformatics **35**, 5191–5198 (2019)
26. Granlund, H., Erkko, P., Eriksson, E., Reitamo, S.: Comparison of cyclosporine and topical betamethasone-17, 21-dipropionate in the treatment of severe chronic hand eczema. Acta Derm. Venereol. **76**(5), 371–6 (1996)
27. Keczkes, K., Frain-Bell, W., Honeyman, A., Sprunt, G.: The effect on adrenal function of treatment of eczema and psoriasis with triamcinolone acetonide. Br. J. Derm.Logy **79**(8–9), 475–86 (1967)
28. Schmied, C., Piletta, P.A., Saurat, J.H.: Treatment of eczema with a mixture of triamcinolone acetonide and retinoic acid a double-blind study. Dermatology **187**(4), 263–7 (1993)
29. Niitani, H., Kobayashi, K.: Cisplatin/carboplatin therapy in nonsmall cell lung cancer. Oncology **49**(Suppl 1), 51 (1992)
30. Saito, H., Shimokata, K., Saka, H., et al.: Phase II study of carboplatin, cisplatin, and vindesine in advanced non-small-cell lung cancer. Cancer Chemother. Pharmacol. **33**, 154–156 (1993)
31. Sculier, J.P., Paesmans, M., Thiriaux, J., Lecomte, J., Bureau, G., Giner, V., Efremidis, A., Lafitte, J.J., Berchier, M.C., Alexopoulos, C.G., Zacharias, C., Mommen, P., Ninane, V., Klastersky, J.: Phase III randomized trial comparing cisplatin and carboplatin with or without ifosfamide in patients with advanced non-small-cell lung cancer. European Lung Cancer Working Party. J. Clin. Oncol. **16**(4), 1388–1396 (1998)
32. Wishart, D.S., et al.: DrugBank: a knowledgebase for drugs, drug actions and drug targets. Nucl. Acids Res. **36**, D901–D906 (2008)
33. Hamosh, A., et al.: Online Mendelian Inheritance in Man (OMIM), a knowledgebase of human genes and genetic disorders. Nucl. Acids Res. **30**, 52–55 (2002)
34. Brown, A.S., Patel, C.J.: A standard database for drug repositioning. Sci. Data **4**, 170029 (2017)
35. Wang, Y., et al.: PubChem: a public information system for analyzing bioactivities of small molecules. Nucl. Acids Res. **37**, W623–W633 (2009)
36. Mitchell, A., et al.: The InterPro protein families database: the classification resource after 15 years. Nucl. Acids Res. **43**, D213–D221 (2015)
37. UniProt, C.: The universal protein resource (UniProt) in 2010. Nucl. Acids Res. **38**, D142–D148 (2010)
38. Kuhn, M., et al.: The SIDER database of drugs and side effects. Nucl. Acids Res. **44**, D1075–D1079 (2015)
39. Calvert, A.H., Harland, S.J., Newell, D.R., Siddik, Z.H., Jones, A.C., McElwain, T.J., Raju, S., Wiltshaw, E., Smith, I.E., Baker, J.M., Peckham, M.J., Harrap, K.R.: (1982) Early clinical studies with cis-diammine-1,1-cyclobutane dicarboxylate platinum II. Cancer Chemother. Pharmacol. **9**(3), 140–7 (1982)

Cognitive States Prediction with KNN and TomekLinks

Quynh Anh Nguyen and Nam Anh Dao

Abstract Detecting the cognitive states of pilots from their electroencephalogram signals is a challenging task and has different applications. At present there are a number of methods available for the cognitive states prediction. However imbalance in data is ultimately the challenge impacting the analysis performance. Here we use condensed nearest neighbors under-sampling technique to recover the class balance of training data before implementing k-nearest neighbors for classification of the electroencephalogram signals. The underlying cognitive states of pilots can be detected from integration of these techniques that include also data pre-processing and dimensionality reduction. Results of experiments for a benchmark database are reported with improved performance.

Keywords EEG · KNN · TomekLinks · PCA · Cognitive states · Imbalance

1 Introduction

With aviation safety statistics spanning from 2018 to 2022 by the International Air Transport Association (IATA) [1], the Aviation Safety Task Force assessed the percentage distribution of incidents as follows: global accidents encompass a range of errors, including those associated with equipment control processes, runway-related issues, and errors stemming from flight control processes. Notably, errors resulting from pilot interventions during flight processes have also been taken into account. The human factor has emerged as a significant element in aviation accident data within IATA, and the importance of the human factor is widely acknowledged in ensuring the safety and performance of the aviation industry. Future advancements in aviation safety are expected to heavily rely on this factor, which encompass physical aspects, cognition, and organizational aspects. Recent studies showed that at least half of aviation accidents are caused by human factor including a range of biological, personal, and medical issues [2, 3].

Q. A. Nguyen · N. A. Dao (✉)
Electric Power University, Hanoi, Vietnam
e-mail: anhdn@epu.edu.vn

N. Hoang Phuong et al. (eds.), *Machine Learning and Other Soft Computing Techniques: Biomedical and Related Applications*, Studies in Systems, Decision and Control 543, https://doi.org/10.1007/978-3-031-63929-6_8

At present there are a number of methods available for cognitive states prediction from electroencephalogram (EEG), galvanic skin response (GSR), electrocardiogram (ECG), and respiration (R) signals to predict pilot cognitive states [4]. These studies are of great help to analyzing psycho-physiological signals to determine pilots' mental states, thereby reducing risks in aviation safety. There are few publications particularly addressing data imbalance data that impact the analysis performance.

In this work, we concentrate especially on predicting pilot cognitive states by learning patterns of electroencephalogram. The condensed nearest neighbors undersampling technique is an important consideration in our study for recovering the class balance of training data. Here, data pre-processing using dimensionality reduction for K-Nearest Neighbors Algorithm (KNN) [5] classification underlying cognitive states of pilots has been explored for a benchmark database Kaggle [6]. The study confirms the value of integration of the KNN classification with imbalance TomekLinks [7, 8] method and this validates high-performance learning model.

The structure of this paper is organized as follows: Sect. 2 reviews studies related to cognitive state identification and data imbalance. Section 3 introduces our method. Section 4 outlines experiments for cognitive state identification and a comparative report. Finally, conclusions are described in Sect. 5.

2 Related Work

There have been many previous studies on cognitive states prediction and so we outline here only on some directly relevant works.

A. Cognitive state identification for reducing commercial aviation fatalities. Gumus et al. [9] focuses on the application of a convolutional neural network (CNN) architecture to identify cognitive states from physiological measurements, including EEG, ECG, GSR, and R data. The study highlights advantage of using the signal without extensive preprocessing which reduces the workload of data preparation for recognition of cognitive states, contributing to the development of real-time cognitive state identification systems. Consider a study conducted by the National Aeronautics and Space Administration regarding attention management in commercial aviation, Harrivel et al. [10] implemented machine learning methods such as Gradient Boosting, Deep Neural Network (DNN), and Random Forest (RF) to classify the cognitive states of pilots. These states include channelized attention, startle/surprise, diverted attention, high workload, and low workload. They combined the outcomes of these three independent methods to provide an overall performance of pilot cognitive states. Mishra et al. [11] developed a model capable of detecting pilots' cognitive states using physiological data and a Support Vector Machine (SVM) as a classifier. They applied a Gaussian Kernel function to address non-linear separability within the data, further optimizing their model.

B. Data imbalance. Jiang and Zhoa [12] proposed a method of addressing the challenge of imbalanced data distribution in seizure detection algorithms based on EEG. The method incorporates hybrid sampling and a cost-sensitive (CS) classifier. A combination of Synthetic Minority Over-sampling Technique (SMOTE) and TomekLinks [7, 8] method were used to resample the imbalanced EEG data. Then, SVM used to classify the resampled EEG feature vectors. In another study, Shchetinin et al. [13] focuses on addressing the issue of deaths caused by cardiovascular diseases. To tackle the challenge of imbalanced datasets, the study utilizes resampling methods such as SMOTE Tomek-Links and SMOTE ENN to enhance the representation of minority classes. Authors applied SVM, RF, Light Gradient Boosting Machine, CNN, Long Short-Term Memory (LSTM), and Bidirectional Long Short-Term Memory (BLSTM) for the classification of irregular heartbeat.

In order to assist healthcare professionals in early-stage screening of Aortic dissection patients with the aim of potentially saving lives, Gao et al. [14] introduced a method known as the Cluster Random under-sampling SMOTE-Tomek Bagging. This approach combines the strengths of Kmeans++ and the SMOTE-Tomek sampling technique to address the issue of imbalanced datasets. The subsequent utilization of the Bagging algorithm aids in the prediction of aortic dissection for patients. The researchers curated a dataset for aortic dissection cases, as well as routine examination data from other cardiovascular patients, sourced from Xiangya hospital, to establish a comprehensive aortic dissection dataset.

Although these methods showed impressive detection performance, the works in [9–11] are not arranged to deal with imbalance, the works in [12–14] are geared towards imbalance data but not designed for the pilots' mental states we aim to achieve. For improving the classifier's performance, the solution presented in this paper is based on imbalance by TomekLinks integrated with KNN and other data processing techniques.

3 The Method

We begin data analysis and feature engineering. We will then look at the imbalance elimination and classification, see Fig. 1.

Data analysis. The EEG data we have are signals containing channels for each person. Identification of the mental states from the data that are time series requires large memory and computing time. By employing a suitable data format, the memory needed for data was decreased, keeping consistency of the data, and improving the computing speed.

The task of noise removal for data was also carried out as there are non-brain activity like electromagnetic interference, physiological artifacts, and amplifier noise. For instance, EEG signals were band-pass filtered and then Independent Component Analysis was applied to avoid the noise's influence on the performance.

Feature engineering. Since the EEG data of each person is a combination of multiple channels, dimensionality reduction is required to decrease the number of features while preserving as much of the significant information as possible. This is achieved by performing principal component analysis (PCA) where high-dimensional data is

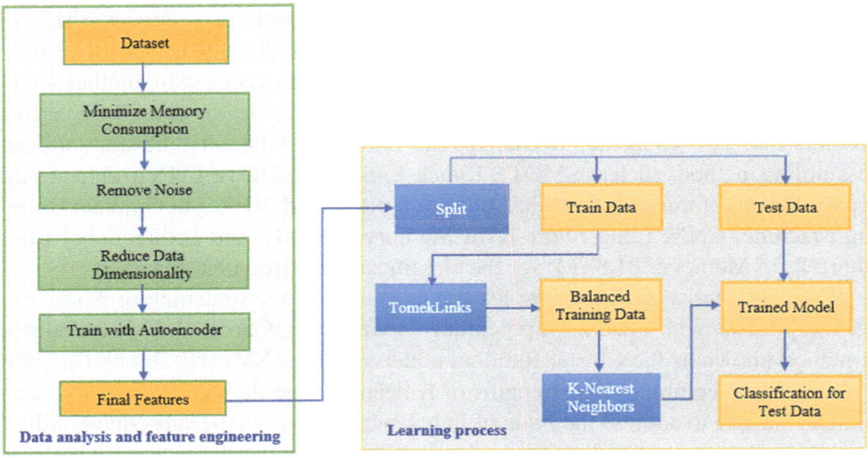

Fig. 1 Cognitive states learning model

transformed into a lower-dimensional space remaining the substance of the original data.

We also compress the EEG data with efficient data representations by autoencoder [18, 19] that actually is an unsupervised learning technique for neural networks. When training the network using back-propagation with three layers including input, a hidden layer for encoding, and output decoding layer, the autoencoder sets the target output values to equal the inputs. The learning process forces the smaller hidden encoding layer to use dimensional reduction to exclude noise and rebuilt the inputs. Hence, this task allows us to produce EEG features with low dimension and noise reduction.

Imbalance elimination. Imbalance occurs when a dataset has too much of samples for some classes (majority) and too little of others (minority). Prior to the imbalance in EEG dataset a balancing method called TomekLinks [7, 8] is proposed to be implemented to the training set. In the method, some samples with its k nearest neighbors from the majority class are selected by a particular rule and then they are removed, decreasing the population of the majority class. The rule to select the pair of observation (say, A and B) requires: (1) The observation A's nearest neighbor is B; (2) The observation B's nearest neighbor is A; (3) Observation A and B belong to a different class. Note, A and B belong to the minority and majority class (or vice versa), accordingly. As the TomekLinks happen, the imbalance in the EEG training dataset is excluded.

Classification. By k-fold cross-validation the effectiveness of our method for the EEG dataset can be estimated. This approach allows each observation in a data sample to appear in a training set or a test set. Thus, each data sample has opportunity to be used in the test set one time and used to train $k - 1$ times. With both features and classes checked and ready for learning, we applied KNN to learn relying on the concepts of similarity and proximity. As suggested in [5], the approach is performed on the EEG dataset by calculating the distance between data samples in order to find the closest data samples from our training set for any test data sample, so that the test data set with the trained model can be classified into metal states.

Indeed, the usefulness of the learning model can be evaluated by performance metrics that are accuracy, precision, recall, F1-score and Matthews correlation coefficient (MCC) (see 1–5 in Appendix). The existing imbalance in the EEG dataset suggests implementation of group mean (GMean) as a metric that attends to maximize the accuracy on each of the classes while holding classes' accuracy balanced (6).

4 Experiments

Having constructed the EEG learning model, we tested it on the benchmark EEG database. Within the above described method, our experiment involves the practical modules, so that EEG signals were analyzed and used for mental stares classification. Algorithm 1 illustrates the algorithm of the learning method, covering data analysis, feature engineering, learning, and performance evaluation.

Algorithm 1: Cognitive States Prediction with KNN and TomekLinks

```
Input  : e – dataset
Output : s – cognitive states
Begin Algorithm
//Data analysis
for each e do
  | e=Reduce memory(e) e=Noise removal(e)
end
//Feature engineering
for each e do
  | e=PCA(e)
  | e=Auto-encoder(e)
end
//Classification by KNN and TomekLinks
(fold, train data, test data)=Split(e)
for each fold do
  | (training data)=Balance with TomekLinks(training data)
  | Train by KNN(train data)
  | Test the model (test data)
end
Calculate accuracy and other metrics
End Algorithm
```

Data analysis. Studies monitoring pilots' psychological states using the EEG signals presented a competition of Reducing Commercial Aviation Fatalities provided by Kaggle challenges [6] scientists to build a model capable for classification of the states. The learning can be useful for warning pilots if they fall into an abnormal

Table 1 The physiological measurements

Variable	Abbreviation	Unit	Description
Electroencephalogram	EEG	Microvolt (μV)	Records electrical activity in the brain
Electrocardiogram	ECG	Microvolt (μV)	Monitors the electrical activity of the heart
Respiration	R	Microvolt (μV)	The action of breathing
Galvanic skin response	GSR	Microvolt (μV)	Refers to the changes in sweat gland activity

state of mind while operating the plane. This is particularly suitable for reducing airplane issues. Closer inspection of the database shows that it encompasses millions of observations and constitutes raw data with noise mentioned in previous section. The physiological data was collected from series of tests managed in a non-flight experiment and outside of a flight simulator. Nine flight crews participated comprising a total of eighteen pilots where a crew is a pair of a captain in seat number 0 and a first officer in seat number 1.

The dataset consists of four groups of physiological measurements indicated in Table 1. All variables are measured at a frequency of 256 Hz. Here the EEG records electrical activity in the brain which detects faddiness in brain waves. The EEG signals are then described by 20 features like eeg o1, eeg o2, eeg fp1. On the other hand, the ECG monitors the electrical activity of the heart while respiration (R)—captures the action of breathing, a measure of the rise and fall of the chest. Galvanic skin response (GSR) refers to the changes in sweat gland activity, a measure of the intensity of our emotional state.

To identify whether a pilot is entered in a troubling physical condition or not, the database shows three statuses available for pilots. The state of channelized attention (CA) involves focusing solely on one task while overlooking all others. This cognitive condition involves omitting other tasks and may lead to unsafe situations. Diverted attention (DA) is a state where one's focus gets redirected by actions associated with decision-making. In aviation operations, diverted attention may causes dangerous situations. Startle/Surprise (SS) is a state that occurs when attention levels are disrupted, involving a surprise or sudden change in the scenario, see Table 2.

Table 2 Cognitive states

Cognitive state	Abbr	Description
Baseline	BL	There is no specific state that occurs
Channelized attention	CA	The state of being focused on one task to the exclusion of all others
Diverted attention	DA	The state of having one's focus diverted related to decision-making
Startle/surprise	SS	When attention levels are disrupted

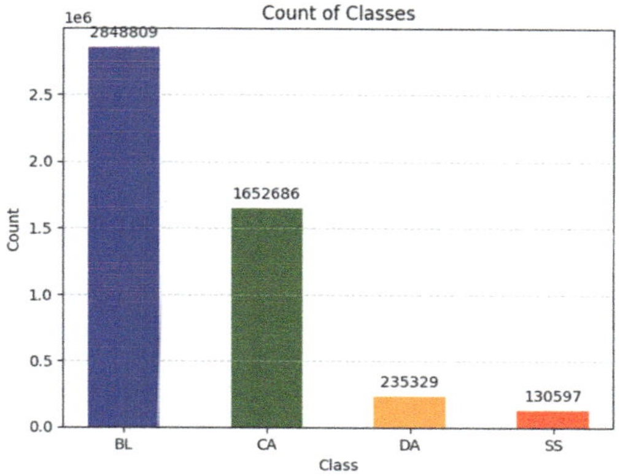

Fig. 2 The distribution of classes in original data. The classes are Baseline (BL), Channelized attention (CA), Diverted attention (DA), Startle/Surprise (SS)

Next, we looked at the distribution of the event variable that means class in our case study. Figure 2 illustrates the data imbalance, indicating a significant disparity in the occurrence frequencies of different events. While some events are extremely frequent, others are notably rare. This shows the imbalance's presence within the dataset.

Feature Engineering. Issue arises when data noise and class imbalance are observed in the dataset during the data analysis. This will be discussed in next subsection. For instance, the stretching of data space was completed when using memory optimization, because of large number of channels and long format of time series. The task reduced memory usage to facilitate further processing to allow the data noise to be decreases to an acceptable level, features were band-pass filtered and then Independent Component Analysis was applied.

With the PCA, the initial data with 20 EEG features now have only two features for the aim of reducing data dimensionality preserving significant information. Then the autoencoders were performed with the data and features were extracted from the hidden layer yielding final features with reduced noise and low dimension.

Imbalance and Classification. Because the distribution of samples into classes in training set varied highly, the data may cause imbalance classification problem where classes SS and DA are minorities. Note that the minority classes SS and DA are more important for our case. Therefore the issue is more sensitive to prediction mistakes for the classes than majority classes.

Attempts to balance the distribution of the classes have particularly centered on more careful selecting the samples with its k nearest neighbors from the majority class and removing them. This aspect of the balancing was performed by the TomekLinks method allowing training data to have equal number of samples for each class.

After splitting and removing imbalance, the samples of training data of EEG features is in a fit state for learning by a similarity measure, which is the basis for determination of neighbours in KNN. The classification method arranges a majority vote between the k most similar samples to a given data sample, making it simple to implement, flexible to choice of similarity measure and natural to handle in our four-class case (*BL, CA, DA* and *SS*). With the classified mental states, pilots can be

Table 3 Metrics with KNN and TomekLinks

K-Folds	Accuracy	Precision	Recall	F1_score	GMean	MCC
Fold 1	0.951	0.949	0.950	0.950	0.861	0.907
Fold 2	0.951	0.949	0.950	0.949	0.859	0.907
Fold 3	0.951	0.949	0.950	0.950	0.861	0.907
Fold 4	0.951	0.949	0.950	0.949	0.859	0.906
Fold 5	0.951	0.949	0.950	0.949	0.862	0.906
Average	0.951	0.949	0.950	0.949	0.860	0.907

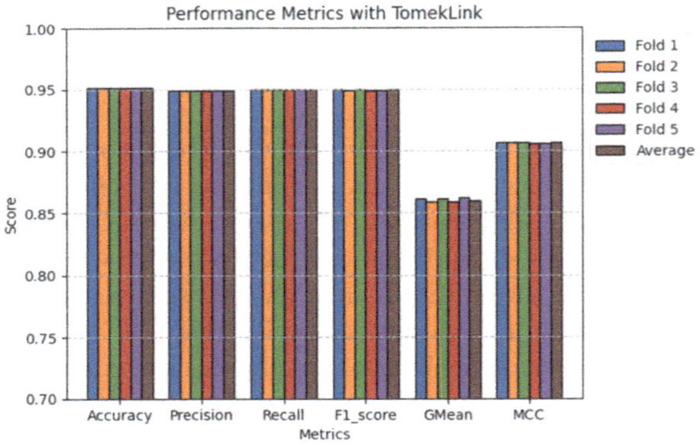

Fig. 3 Performance metrics

alerted before they enter dangerous states, enhancing aviation safety and reducing accidents.

Discussion. The results of the experiments are described in Table 3 and displayed in Fig. 3. Following equation (1), the accuracy of 0.951 shows how often our learning model exactly predicts the state by dividing the number of correct predictions by the total number of test samples. It is 0.86 measured by Gmean (6) for imbalance data because the metric tend to raise the accuracy on each of the classes. The metrics for KNN with TomekLinks are 0.949 for precision (2), 0.950 for recall (3), 0.949 for F1_score (4) and 0.907 for the Matthews correlation coefficient (5).

As we mentioned in Sect. 1, the problem of cognitive states prediction from electroencephalogram has received very strong attention in recent research for reducing risks in aviation safety. Publications of analyzing psycho-physiological signals using different methods are listed in the Table 4.

With the same dataset from Kaggle [6], Gumus and Saylam [9] in 2023 implemented CNN with the SoftMax classifier giving the accuracy of 0.89. In relation for determining the pilot's cognitive states, Lin [16] applied Random Forest classification having accuracy by 0.897 with all features including EEG, ECG, R, GSR, crew and seat. The classification of pilots' mental states associated with human attention

Table 4 Predictive models

No	Method	Database	Accuracy
1	CNN with the SoftMax classifier [9]	Kaggle2019 [6]	0.890
2	Random Forest, all features [16]	Kaggle2019 [6]	0.897
3	SMOTE, Cosine Similarity, XGBoost [20]	AHPLS dataset [20]	0.916
4	K-means, SMOTE [21]	iNeuro database [21]	0.926
5	Extreme Gradient Boosting and k-means [17]	Kaggle2019 [6]	0.942
6	Our model	Kaggle2019 [6]	0.951

was addressed by Alreshidi et al. [20] using an imbalance physiological dataset. Here, the combination of SMOTE and cosine similarity produced accuracy of 0.916. Hu et al. [21] introduced the K-means SMOTE for seizure detection, achieving accuracy of 0.926. Note that the last two works used different databases for their experiments.

The combination of the K-means clustering method and the Extreme Gradient Boosting classifier in our previous paper [17] achieved accuracy of 0.942, whereas the KNN in the integration with imbalance method of TomekLinks allowed to have accuracy of 0.951 in this work. Quantitatively, the imbalance technique can eliminate the number of majority samples in the dataset for training. Qualitative performance evaluation is only improved for dataset having imbalance. This is usually suggested for data analysis with the absence of class balance. Our solution is based on the imbalance analysis for time series data, it is not limited to the pilots' mental states estimation and can be applied for other cases.

5 Conclusions

The method provided in this paper outlines an imbalance based model for learning patterns of electroencephalogram. The objective is to improve the prediction of the pilot's cognitive state which helps to reduce the aviation accidents. The method focused on three main techniques. First, we optimized the memory usage, eliminated noisy data and employed PCA to reduce data dimensionality. Second, the TomekLink which is a condensed nearest neighbors under-sampling technique was applied to deal with imbalance issue. Finally, the implementation of KNN enabled the classification of underlying cognitive states. Experiments demonstrated the value of combination of the KNN classification with imbalance TomekLinks and validated high-performance learning model. The proposed method can be applied to electroencephalogram signals as well as to other time-varying bioelectrical signals for monitoring.

Acknowledgements The paper is the result of the scientific research project at the Electric Power University in 2023.

6 Appendix

Given True Positives (TP), True Negatives (TN), False Positives (FP), False Negatives (FN), there are the following metrics:

$$Accuracy = \frac{TP + TN}{TP + TN + FP + FN}, \tag{1}$$

$$Precision = \frac{TP}{TP + FP}, \tag{2}$$

$$Recall = \frac{TP}{TP + FN}, \tag{3}$$

$$F1\_score = 2 * \frac{Precision * Recall}{Precision + Recall}, \tag{4}$$

$$MCC = \frac{TN \times TP - FN \times FP}{\sqrt{(TP + FP)(TP + FN)(TN + FP)(TN + FN)}}, \tag{5}$$

$$GMean = \sqrt{Sensitivity \cdot Specificity}, \tag{6}$$

$$Sensitivity = \frac{TP}{TP + FP}, \tag{7}$$

$$Specificity = \frac{TN}{TP + FP}. \tag{8}$$

References

1. IATA Annual Safety Report-2022. https://www.iata.org/contentassets/a8e49941e8824a058fee3f5ae0c005d9/safety-report-executive-and-safety-overview.pdf
2. Zavila, O., Mach, O., Bauer, M.: Methods for the identification and analysis of human errors in current military aviation. In: International Conference on Military Technologies, (IEEE, 2021), pp. 1–6 (2021). https://doi.org/10.1109/ICMT52455.2021.9502798
3. İnan, T.T., İnan, N.G.: The analysis of fatal aviation accidents more than 100 dead passengers: an application of machine learning. OPSEARCH 1–19 (2022). https://doi.org/10.1007/s12597-022-00585-1
4. Harrivel, A.R., Stephens, C.L., Milletich, R.J., Heinich, C.M., Last, M.C., Napoli, N.J., Abraham, N., Prinzel, L.J., Motter, M.A., Pope, A.T.: Prediction of cognitive states during flight simulation using multimodal psychophysiological sensing. In: AIAA Information Systems, AIAA Infotech@Aerospace (2017). https://doi.org/10.2514/6.2017-1135
5. Campos, G.O., Zimek, A., Sander, J., Campello, R.J.G.B., Micenková, B., Schubert, E., Assent, I., Houle, M.E.: On the evaluation of unsupervised outlier detection: measures, datasets, and an empirical study. Data Min. Knowl. Discov. **30**(4), 891–927 (2016). https://doi.org/10.1007/s10618-015-0444-8. ISSN 1384-5810. S2CID 1952214

6. Hamilton, B.A., Kaggle.: Reducing commercial aviation fatalities. https://www.kaggle.com/c/reducing-commercial-aviation-fatalities (2019). Accessed 22 Apr 2020
7. Tomek, I.: An experiment with the edited nearest-neighbor rule. IEEE Trans. Syst. Man Cybern. **SMC-6**(6), 448–452 (1976). https://doi.org/10.1109/TSMC.1976.4309523.
8. Kamaladevi, M., Venkataraman, V., Sekar, K.R.: Tomek link undersampling with stacked ensemble classifier for imbalanced data classification. Ann. Rom. Soc. Cell Biol. 2182 (2021). Accessed from http://www.annalsofrscb.ro/index.php/journal/article/view/2751
9. Gumus, F., Saylam, R.: Prevention of aviation accidents with prediction of cognitive states. In: 10th International Conference on Recent Advances in Air and Space Technologies (RAST), Istanbul, Turkiye, pp. 1–4 (2023). https://doi.org/10.1109/RAST57548.2023.10197937.
10. Harrivel, A.R., Liles, C., Stephens, C.L., Ellis, K.K., Prinzel, L.J., Pope, A.T.: Psychophysiological sensing and state classification for attention management in commercial aviation, pp. 1–8 (2016)
11. Mishra, A., Shrivastava, K.K., Anto, A.B., Quadir, N.A.: Reducing commercial aviation fatalities using support vector machines. In: 2019 International Conference on Smart Systems and Inventive Technology (ICSSIT), pp. 360–364. IEEE (2019)
12. Jiang, Z., Zhao, W.: Fusion algorithm for imbalanced EEG data processing in seizure detection. Seizure Eur. J. Epilepsy 207–211 (2021). https://doi.org/10.1016/j.seizure.2021.06.023
13. Shchetinin, E.Y., Glushkova, A.G.: Arrhythmia detection using resampling and deep learning methods on unbalanced data. Comput. Opt. **46**(6), (2022). https://doi.org/10.18287/2412-6179-CO-1112
14. Gao, Y., Wang, M., Zhang, G., Zhou, L., Luo, J., Liu, L.: Cluster-based ensemble learning model for aortic dissection screening. Int. J. Environ. Res. Public Health **19**(9), 5657 (2022)
15. Kubat, M., Matwin, S.: Addressing the curse of imbalanced training sets: one-sided selection. In: Proceedings of the Fourteenth International Conference in Machine Learning, pp. 179–186 (1997)
16. Lin, Y.-C.: Reducing aviation fatalities by monitoring pilots' cognitive states using psychophysiological measurements. Thesis, Naval Postgraduate School, Monterey, California (2021)
17. Dao, N.A., Nguyen, Q.A.: Mental states detection by extreme gradient boosting and k-means. In: The Fourth International Conference on Artificial Intelligence and Computational Intelligence (2023). https://doi.org/10.1007/978-3-031-29447-12
18. Pascal, V., Hugo, L.: Stacked denoising autoencoders: learning useful representations in a deep network with a local denoising criterion. J. Mach. Learn. Res. **11**, 3371–3408 (2010)
19. Bank, D., Koenigstein, N., Giryes, R.: Autoencoders (2021). arXiv:2003.05991v2. https://doi.org/10.48550/arXiv.2003.05991
20. Alreshidi, I., Yadav, S., Moulitsas, I., Jenkins, K.: A comprehensive analysis of machine learning and deep learning models for identifying pilots' mental states from imbalanced physiological data. In: AIAA 2023-4529. AIAA AVIATION 2023 Forum (2023). https://doi.org/10.2514/6.2023-4529
21. Hu, D., Cao, J., Lai, X., Liu, J., Wang, S., Ding, Y.: Epileptic signal classification based on synthetic minority oversampling and blending algorithm. IEEE Trans. Cogn. Dev. Syst. **13**(2), 368–382 (2021). https://doi.org/10.1109/TCDS.2020.3009020

Health Digital Twins with Clinical Decision Support and Medical Imaging

Moritz Grob, Andrea Rappelsberger, and Klaus-Peter Adlassnig

Abstract As the concept of digital twins as virtual representations of physical entities is rapidly gaining popularity in healthcare, this paper studies the feasibility of incorporating clinical decision support (CDS), clinical data, and medical imaging into health digital twins (HDTs). A HDT is visualized in a web application, health data are stored in a FHIR-based electronic health record, computed tomography images are stored in a rudimentary DICOM-based picture archiving and communication system. An Arden-Syntax-based CDS system consisting of an interpretation and an alert service is connected. The prototype focuses on interoperability of these components. The study confirms the feasibility of CDS integration into HDTs and provides insight into possibilities for further expansion.

1 Introduction

In the last decades, the healthcare industry has witnessed significant advances in technology-driven solutions aimed at enhancing patient care, diagnosis, and treatment.

Clinical Decision Support. One such advancement can be observed in computerized clinical decision support systems (CDSSs), which over the past 25 years have evolved from simple stand-alone systems lacking meaningful interoperability to complex, integrated platforms [16].

Challenges remain regarding the universal implementation of CDSSs in clinical practice, such as a quickly evolving technological landscape leading to the need of processing vast amounts of various clinical data, the persistent difficulty in establishing standardized intra-organizational processes for knowledge engineering and

M. Grob · A. Rappelsberger · K.-P. Adlassnig (✉)
Medical University of Vienna, Center for Medical Data Science, Institute of Artificial Intelligence, Spitalgasse 23, 1090 Vienna, Austria
e-mail: klaus-peter.adlassnig@meduniwien.ac.at; kpa@medexter.com

M. Grob · K.-P. Adlassnig
Medexter Healthcare, Borschkegasse 7/5, 1090 Vienna, Austria

N. Hoang Phuong et al. (eds.), *Machine Learning and Other Soft Computing Techniques: Biomedical and Related Applications*, Studies in Systems, Decision and Control 543, https://doi.org/10.1007/978-3-031-63929-6_9

management, and the need to further integrate seamlessly and efficiently into clinical workflows [16]. Nonetheless, a 2020 meta-analysis of controlled clinical trials by Kwan et al. [12] found that the use of CDSSs led on average to an overall absolute improvement in the proportion of patients receiving the desired care by 5.8% (CI_{95} 4.0% to 7.6%), albeit with significant heterogeneity between studies (I^2 = 76%). While this improvement may not seem revolutionary, no data regarding the speed at which the desired care had been delivered was collected. Furthermore, it is predicted that adoption as well as efficacy of CDSSs will continue to rise in the coming years as a result of continuing improvements in computational capabilities and knowledge discovery [16].

Other factors contributing to the predicted further rise in CDSSs' relevance to clinical practice is the ongoing adoption of Health Level Seven's (HL7) Fast Healthcare Interoperability Resources (FHIR) as a unified clinical data model as well as the evolution of standards representing clinical knowledge and clinical reasoning—like for instance the Clinical Quality Language (CQL) and Arden Syntax (both normative standards by HL7)—and their interoperability with the FHIR data model [19].

Health Digital Twins. Another concept rapidly gaining traction is that of the health digital twin (HDT). Originally emanating from the manufacturing industry, developments in information and communications technology have given rise to increasing research on how to apply digital twins as virtual representations of real-world entities to the healthcare sector [20]. While the classical approach to healthcare consists primarily of one-size-fits-all guidelines, tailored to a hypothetical average patient, HDTs bear the potential to facilitate a more individualized, patient-centered healthcare delivery [14]. A variety of research projects on utilization of digital twins in clinical care are underway [11, 14, 20, 21]. However, they tend to focus on a narrow field of medicine such as personalized disease management in multiple sclerosis [21]. Thus far, a more generalized approach is lacking.

Health Digital Twins and Clinical Decision Support. This study presents a prototype web application integrating the concept of the HDT with generalized clinical data stored in a FHIR-conformant electronic health record (EHR), medical imaging data stored in a picture archiving and communication system (PACS), and CDSSs.

Since a high-fidelity full-body HDT may be far from achievable due to technological limitations, the HDT aims to be a proof of concept of a full-body low-fidelity proto-HDT [11]. Low-fidelity in this context means that the underlying data are not updated in real time—e.g., by use of internet of things sensors—but only when new laboratory test results or medical imaging data of the patient are available and the medical professional actively uses the application. The prefix proto- denotes that the current implementation serves as an archetype for building more complex, individualized HDTs.

Furthermore, the prototype integrates state-of-the-art medical imaging visualization functionality. The application is intended to integrate into a hospital information system as depicted in Fig. 1. This integration is possible by adherence to industry-standard interfaces—namely FHIR as a standard data model as well as transmission of clinical data via the FHIR representational state transfer (REST) application

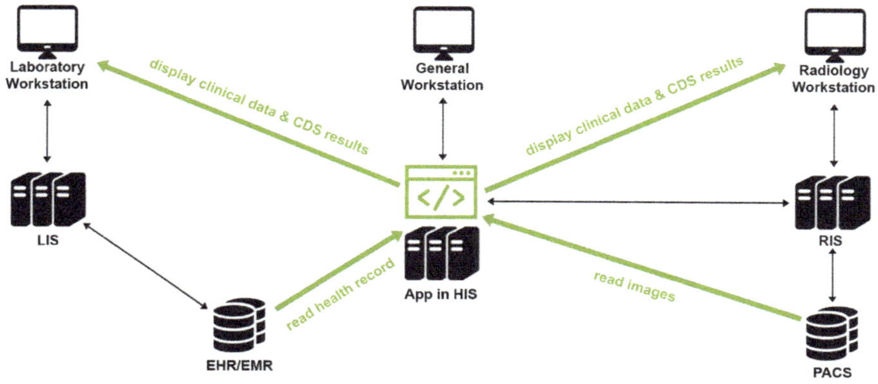

Fig. 1 Diagram of the proposed application's integration into a hospital information system (HIS). It reads the patient's clinical data—entered for example via a laboratory information system (LIS)—from an electronic health record (EHR) or electronic medical record (EMR) while reading image data from the picture archiving and communications system (PACS) without having to connect to the radiology information system (RIS). These images can then be displayed together with the clinical data and clinical decision support (CDS) results on any workstation in the hospital environment

programming interface (API) and Digital Imaging and Communications in Medicine (DICOM) for storage and transmission of medical imaging data.

2 Methods

A literature research concerning the state-of-the-art of HDTs was conducted using the MEDLINE/PubMed database by the United States National Institutes of Health as well as Google Scholar and ResearchGate.

Next, a web application was implemented according to the Unified Modeling Language (UML) component diagram depicted in Fig. 2. The frontend was developed

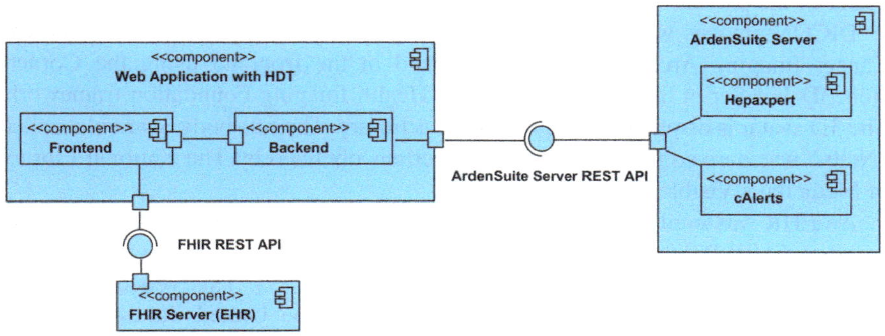

Fig. 2 UML 2.5 component diagram of the prototype system

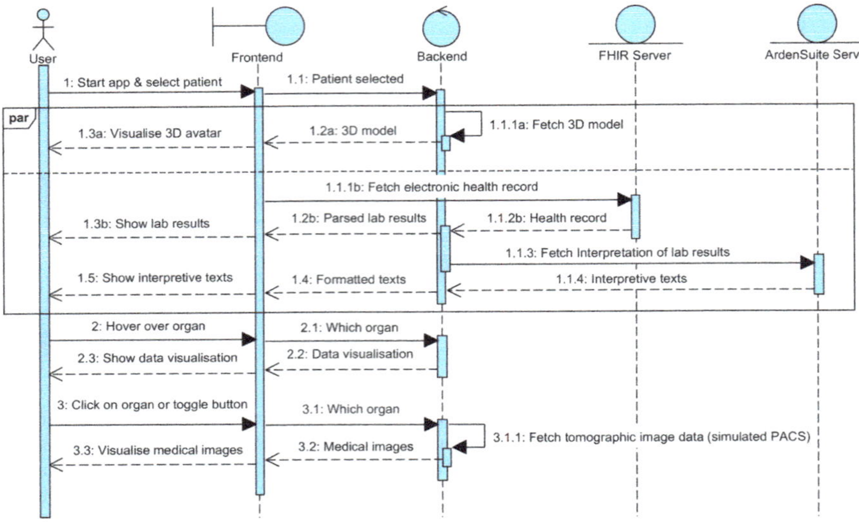

Fig. 3 UML 2.5 sequence diagram of the application where the frontend is shown as the system's boundary towards the user, the backend as controller, and the FHIR and ArdenSuite servers as entities

using the Angular web development platform, the Angular Material user interface component library, the Tailwind Cascading Style Sheets (CSS) framework, and the WebStorm integrated development environment (IDE). Version control was achieved via GitLab. The application flow is depicted in the form of a UML sequence diagram in Fig. 3.

The backend for the web application was developed using the Express web framework for Node.js and the WebStorm IDE. It functions as a rudimentary PACS by storing DICOM P10 instances and providing them to the frontend via the Web Access to DICOM Objects via Uniform Resource Identifiers (WADO-URI) communication mode described in Chapter 6 of part 18 of the DICOM standard [18]. Furthermore, the backend statically serves a three-dimensional (3D) model acting as a full-body avatar of the patient.

DICOM images were obtained from the HCC-TACE-Seg dataset [17] via The Cancer Imaging Archive and are visualized in the frontend using the Cornerstone3D JavaScript library from the Open Health Imaging Foundation framework. The 3D avatar is displayed using the Three.js library. The full-body 3D model called "Nelly" was derived from digitized cryosections obtained by The National Library of Medicine's Visible Human Project [15].

An EHR was implemented using a Docker image of the HAPI FHIR JPA server based on FHIR Release 4, the first release containing normative elements approved by the American National Standards Institute (ANSI) [7]. This version of FHIR is the data model and API standard mandated by the United States 21st Century Cures Act [1]. Furthermore, FHIR is to be adopted as a common standard in the European Union as decided by the European Union eHealth Network in early 2023 [2, 13].

Exemplary FHIR resources—i.e, a Patient resource containing demographic data and Observation resources containing clinical data—were created using the HAPI FHIR Java library. For semantic interoperability, all information in the resources is based on the reference terminologies Logical Observation Identifiers Names and Codes (LOINC), Systematized Nomenclature of Medicine Clinical Terms (SNOMED CT) and Unified Code for Units of Measure (UCUM).

The CDSS was implemented using the ArdenSuite CDS Platform [5], which is based on version 2.9 of the HL7 Arden Syntax [4, 8, 10, 19]. In the current prototype, it consists of two separate subsystems: *Hepaxpert* [3] for the automated interpretation of hepatitis serology laboratory test results as well as *cAlerts* [22] for providing context-sensitive clinical alerting. While *cAlerts* consists of multiple different packages, currently only one regarding myocardial infarctions via monitoring of blood levels of the cardiac troponin enzyme is implemented in the prototype application.

Since the application is in the proof-of-concept stage, merely the clinical data required for the CDSS are retrieved from—as well as initially saved in—the EHR, namely FHIR Observation resources containing laboratory test results regarding hepatitis and cardiac troponin serology. Nonetheless, it is designed to facilitate generalization via expansion of the CDS modules or just by acquisition and display of a wider range of clinical data.

3 Results

The proposed prototype was established as described in Sect. 2 and the components were successfully interconnected. Upon starting the application, the Visible Human Project HDT is visualized showing its anatomical parts separately and colored for easier differentiation. The patient's clinical data is fetched from the FHIR-based EHR, parsed, and visualized in tabular form besides the avatar upon a click on the heading "Laboratory Test Results", as visualized on the left side of Fig. 5. At the same time, the CDSS is queried with the parsed clinical data, and alert messages derived from the results are displayed on the respective organs as depicted in the left part of Fig. 4. If interested in a comprehensive summary of the laboratory test results' clinical significance, interpretive texts generated by the CDSS can be displayed in the rightmost column of the screen with a click on the heading "Interpretive Texts",

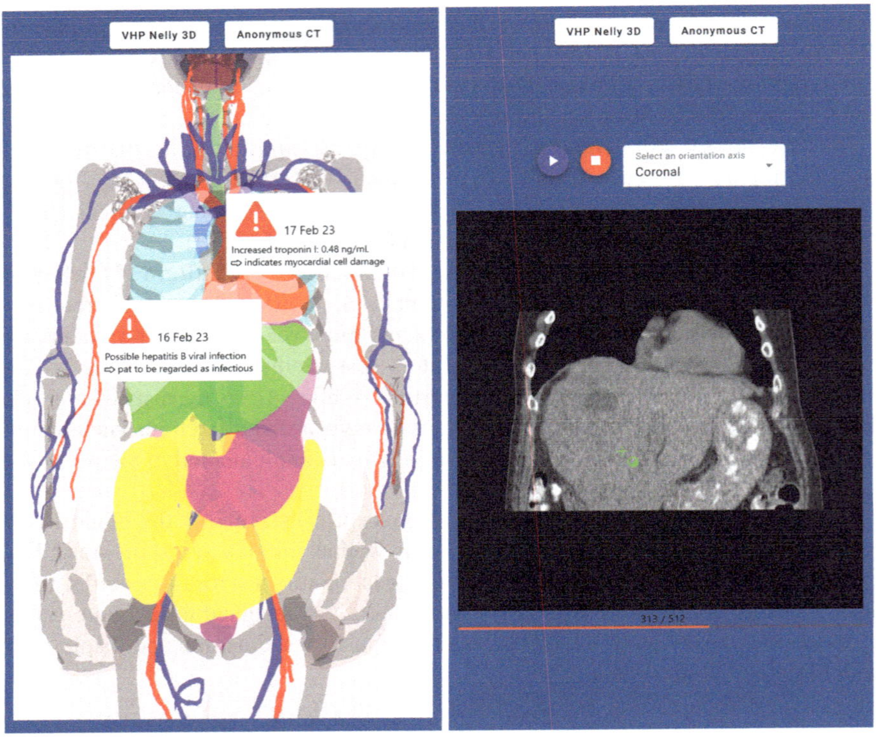

Fig. 4 Screenshots of the HDT component of the application. The full-body 3D avatar with the CDS alert messages visualized as tooltips on top of their respective organs is depicted on the left. Upon clicking on an organ or the heading "Anonymous CT", the avatar is replaced with a DICOM viewer showing recent images of the patient, in this case of the patient's liver. The DICOM viewer is depicted on the right and allows for manipulation of the window width and window center with the cursor to highlight the contrast of different tissues or materials, scrolling through a stack of slices with the mouse wheel or animating the stack scroll via the play and stop buttons, and multiplanar reconstruction via the choice box above the images

as can be seen in the right column of Fig. 5. Clicking on an organ of interest leads to the most recent diagnostic images to be fetched from the PACS and displayed instead of the full-body avatar. This is depicted in the right part of Fig. 4, where a computed tomography (CT) series of the patient's liver consisting of 97 instances (i.e., slices) is visualized. Incorporation of the novel Cornerstone3D library for the DICOM viewer facilitates further inspection of the patient's anatomy by incorporating multiplanar reconstruction, meaning that the application has the ability to calculate the volume between the slices, enabling the series to be viewed from different orientation axes. Furthermore, the window width and window center of the displayed CT images can be easily adjusted with the mouse, allowing for different types of tissue to be inspected in sufficient contrast.

Fig. 5 Screenshot of the CDS component of the application. Laboratory test results parsed from FHIR Observation resources are displayed in the left column, while the CDS messages obtained from the ArdenSuite server are displayed in the right column. Both these columns are displayed to the right of the HDT component depicted in Fig. 4. Laboratory test results are color-coded for better readability, where "negative" is colored in green, "borderline" in orange, and "positive" as well as values leading to a clinical alert are colored in red

4 Discussion

A prototype application integrating HDTs, EHRs, PACS, and CDSSs was presented, confirming the feasibility of combining these concepts. However, due to the archetypal nature of the application, there remain several possibilities for future work. For instance, it would be possible to expand the DICOM viewer functionality to perform volumetric rendering as well as the already implemented multiplanar reconstruction. This would enable display of a complete 3D model of the imaged body region. Moreover, DICOM segmentation objects could be visualized on top of the CT images. In conjunction with volumetric rendering, this feature could make the full-body avatar created from cryosections obsolete in the future, instead creating an HDT out of the patient's latest tomographic images.

Current versions of the Arden Syntax do not standardize data access from external sources. Instead, site-specific implementations are allowed and expressed within curly braces ({ }) in medical logic modules (MLMs). Thus, sharing of clinical knowledge represented in MLMs reliant on accessing external databases requires manual adaption of each MLM, a challenge dubbed the curly braces problem [4, 10]. This problem is avoided in the prototype by the application parsing the clinical data fetched from the EHR, converting the contents of the FHIR Observation resources into data types clearly defined by the standard, and querying the ArdenSuite server with those objects. The incorporation of FHIR as a unified data model in Arden Syntax version 3.0 [9] will allow for MLMs directly accessing data from FHIR resources [10, 19], thus facilitating substantial architectural simplification of the proposed application.

Integration of live sensor data is conceivable, increasing the fidelity of the proposed HDT and thereby broadening its field of application to include intensive or surgical care, as for example detailed in [14]. However, the collection and analysis of "big" data was outside the scope of this feasibility study.

The current prototype relies on a medical professional to actively engage in the application for querying the CDSS. A way of further integrating the application into clinical workflows would be the inclusion of CDS Hooks as a mechanism to automatically query the ArdenSuite server for CDS whenever a new laboratory result is added to the EHR [4, 6].

In order to allow for the HDT to be eventually used in clinical practice, execution of a qualitative usability study and subsequent quantitative validation of the proposed system will be the inevitable next steps in development.

5 Conclusion

The study confirms the feasibility of integrating clinically proven CDSS into a full-body HDT and demonstrates the clear potential for promoting patient-centered care as well as expediting clinical workflows. Nevertheless, more research is needed to adequately assess clinical usefulness as well as the system's impact quality of care.

References

1. 21st century cures act: interoperability, information blocking, and the ONC health IT certification program. Fed. Regist. **85**(85), 25642–25961 (2020). https://www.federalregister.gov/d/2020-07419
2. 23rd eHealth Network Meeting–Minutes of Meeting. European Commission, Directorate-General for Health and Food Safety, Brussels, Belgium (2023). https://health.ec.europa.eu/system/files/2023-05/ehealth_20230330_sr_en.pdf
3. Adlassnig, K.-P., Gawrylkowicz, J., Rappelsberger, A.: Interpretive knowledge engines: a new service for pathologists to bridge the gap to clinicians. In: Abstracts 31st European Congress

of Clinical Microbiology & Infectious Diseases (ECCMID), Vienna, Austria, 9–12 July 2021. European Society of Clinical Microbiology and Infectious Diseases (ESCMID) eLibrary: 516
4. Adlassnig, K.-P., Haug, P., Jenders, R.A.: Arden Syntax: then, now, and in the future. Artif. Intell. Med. **92**, 1–6 (2018). https://doi.org/10.1016/j.artmed.2018.09.001
5. ArdenSuite CDS Platform. Medexter Healthcare GmbH, Vienna, AT. https://www.medexter.com/products-and-services/ardensuite
6. de Bruin, J.S., Rappelsberger, A., Adlassnig, K.-P., Gawrylkowicz, J.: Exploring Methods of Implementing Arden Syntax for CDS Hooks. In: Schreier, G., Hayn, D., Eggerth, A. (eds.) dHealth 2020. Studies in Health Technology and Informatics, vol. 271, pp. 191–198. IOS Press, Amsterdam, Netherlands (2020). https://doi.org/10.3233/shti200096
7. Fast Healthcare Interoperability Resources (FHIR) Standard, Release 4 (Technical Correction #1). Health Level Seven International, Ann Arbor, MI, USA (2019). https://hl7.org/fhir/R4
8. Health Level Seven Arden Syntax for Medical Logic Systems Standard, Version 2.9. Health Level Seven International, Ann Arbor, MI, USA (2013). https://www.hl7.org/implement/standards/product_brief.cfm?product_id=290
9. Health Level Seven Arden Syntax for Medical Logic Systems Standard, Version 3.0 for ballot. Health Level Seven International, Ann Arbor, MI, USA (2023). https://confluence.hl7.org/download/attachments/120750596/2023-10-04_Arden_Syntax_3.0_STU_for_publication.pdf
10. Jenders, R.A., Adlassnig, K.-P., Fehre, K., Haug, P.: Evolution of the Arden Syntax: key technical issues from the standards development organization perspective. Artif. Intell. Med. **92**, 10–14 (2018). https://doi.org/10.1016/j.artmed.2016.08.001
11. Kamel Boulos, M.N., Zhang, P.: Digital Twins: From Personalised Medicine to Precision Public Health. J. Pers. Med. **11**(8), 745 (2021). https://doi.org/10.3390/jpm11080745
12. Kwan, J.L., Lo, L., Ferguson, J., Goldberg, H., Diaz-Martinez, J.P., Tomlinson, G., Grimshaw, J.M., Shojania, K.G.: Computerised clinical decision support systems and absolute improvements in care: meta-analysis of controlled clinical trials. BMJ **370**, m3216 (2020). https://doi.org/10.1136/bmj.m3216
13. Lau, T.: Gesundheitsdaten: Eine Sprache für Europa. Deutsches Ärzteblatt International **120**(16), A-703–A-704 (2023). https://www.aerzteblatt.de/int/article.asp?id=230797
14. Lonsdale, H., Gray, G.M., Ahumada, L.M., Yates, H.M., Varughese, A., Rehman, M.A.: The perioperative human digital twin. Anesth. & Analg. **134**(4), 885–892 (2022). https://doi.org/10.1213/ane.0000000000005916
15. Makarov, S.N., Noetscher, G.M., Yanamadala, J., Piazza, M.W., Louie, S., Prokop, A., Nazarian, A., Nummenmaa, A.: Virtual human models for electromagnetic studies and their applications. IEEE Rev. Biomed. Eng. **10**, 95–121 (2017). https://doi.org/10.1109/rbme.2017.2722420
16. Middleton, B., Sittig, D.F., Wright, A.: Clinical decision support: a 25 year retrospective and a 25 Year vision. Yearb. Med. Inform. (Suppl. 1), S103–S116 (2016). https://doi.org/10.15265/iys-2016-s034
17. Moawad, A.W., Fuentes, D., Morshid, A., Khalaf, A.M., Elmohr, M.M., Abusaif, A., Hazle, J.D., Kaseb, A.O., Hassan, M., Mahvash, A., Szklaruk, J., Qayyom, A., Elsayes, K.: Multimodality annotated HCC cases with and without advanced imaging segmentation. The Cancer Imaging Archive (2021). https://doi.org/10.7937/tcia.5fna-0924
18. NEMA PS3 / ISO 12052, Digital Imaging and Communications in Medicine (DICOM) Standard. National Electrical Manufacturers Association, Rosslyn, VA, USA. https://www.dicomstandard.org/
19. Strasberg, H.R., Rhodes, B., Del Fiol, G., Jenders, R.A., Haug, P.J., Kawamoto, K.: Contemporary clinical decision support standards using Health Level Seven International Fast Healthcare Interoperability Resources. J. Am. Med. Inform. Assoc. **28**(8), 1796–1806 (2021). https://doi.org/10.1093/jamia/ocab070
20. Sun, T., He, X., Li, Z.: Digital twin in healthcare: recent updates and challenges. Digit. Health **9**, 20552076221149652 (2023). https://doi.org/10.1177/20552076221149651

21. Voigt, I., Inojosa, H., Dillenseger, A., Haase, R., Akgün, K., Ziemssen, T.: Digital twins for multiple sclerosis. Front. Immunol. **12**, 669811 (2021). https://doi.org/10.3389/fimmu.2021.669811
22. Zeckl, J., Adlassnig, K., Fossler, R., Blacky, A., de Bruin, J.S., Koller, W., Rappelsberger, A., Adlassnig, K.-P.: Context-sensitive clinical alert packages written in Arden Syntax. In: Gundlapalli, A.V., Jaulent, M.-C., Zhao, D. (eds.) MEDINFO 2017. Studies in Health Technology and Informatics, vol. 245, pp. 1190–1194. IOS Press, Amsterdam, Netherlands (2017). https://doi.org/10.3233/978-1-61499-830-3-1190

Promoting STEM-Integrated Learning Through Engineering Design: High School Students' Automatic Hand Washers

Linh Thuy Thi Pham, Hai Thanh Nguyen, Long Diep, Hoa Huu Nguyen, and Ngoc Cam Thi Tran

Abstract In the current education world, STEM (Science, Technology, Engineering, and Math) is one of the leading methods of interest. The STEM education model is a harmonious combination of the fields of Science, Technology, Engineering, and Mathematics. Instead of existing separately, this knowledge interacts, intertwines, and complements each other through practice activities and lessons. It helps students deepen their understanding of theoretical concepts and encourages them to apply that knowledge in practice by creating practical, functional products. A STEM lesson built on the engineering design process will include eight steps divided into five main activities that we put into the research base, including problem identification, research background knowledge, proposing solutions to meet the set criteria, presenting and discussing design options, product manufacturing, and evaluation testing during manufacturing; product presentation and completion. In this article, we offer the technical design process of the automatic hand washing machine model. We experimented with two classes, including 70 students at An Khanh High School, Can Tho, Vietnam, and the performance evaluation of activities based on the design rubric showed positive results. Two products have been selected for direct use in schools in Can Tho City, Vietnam, during the post-Covid-19 recovery period.

L. T. Thi Pham · L. Diep
Can Tho University of Technology, Can Tho, Vietnam
e-mail: dlong@ctuet.edu.vn

H. T. Nguyen (✉) · H. H. Nguyen
Can Tho University, Can Tho, Vietnam
e-mail: nthai.cit@ctu.edu.vn

H. H. Nguyen
e-mail: nhhoa@ctu.edu.vn

N. C. Thi Tran
Dong Thap University, Dong Thap, Vietnam

N. Hoang Phuong et al. (eds.), *Machine Learning and Other Soft Computing Techniques: Biomedical and Related Applications*, Studies in Systems, Decision and Control 543, https://doi.org/10.1007/978-3-031-63929-6_10

1 Introduction

STEM has emerged as a leading approach in today's education sector, attracting considerable attention due to its innovative and inclusive nature. STEM education integrates distinct subjects, including science, technology, engineering, and math, into a cohesive framework that fosters a deep understanding of the subjects and forms essential skills, such as problem-solving, creative thinking, and collaboration.

As the world is driven by technology and the need for a technically qualified workforce continues to grow, STEM education becomes all the more critical because of its potential to cultivate the necessary skills and mindsets. According to research [1], the importance of STEM education has been demonstrated through compelling evidence, including both opportunities and challenges in promoting STEM education.

According to a study conducted by [2, 3], students are increasingly interested in STEM education due to its hands-on nature and practical relevance. This research shows that students participating in STEM-related activities exhibit higher enthusiasm, interest, and participation levels than traditional learning methods. Furthermore, a study [4] indicates that students actively involved in STEM education tend to outperform their peers on various learning assessments, indicating the effectiveness of the approach method in improving overall learning outcomes. Findings suggested that STEM activities can be implemented to improve students' science process skills, STEM career interests, motivation, and views about STEM education [5].

A standardized and practical STEM competency assessment tool for STEM education is vital in guiding future development. We use the rubric tool to provide evaluation criteria for each study activity. A rubric [6] is a guide that includes specific criteria arranged to different degrees to evaluate the quality or effectiveness of a product or process. The rubric helps to define the evaluation criteria objectively and transparently. Each rubric level usually has a specific description of what is expected at each level.

In this study, we discuss the engineering process of designing automatic hand washing machines in the STEM framework. We aim to not only create a functional device, but also demonstrate how STEM education can lead to more successful educational outcomes. Our contributions to this topic are as follows:

- We develop specific assessment criteria for each activity in STEM, assessing not only the end product but also the process, promoting a more holistic assessment of student learning.
- Demonstrate the benefits of STEM education by emphasising its impact on learning outcomes. We analyzed the learning outcomes of students participating in the automatic hand washing machine design process, showing its superiority over traditional teaching methods.

The rest of this study is organized as follows. Section 2 introduces the main related works. Next, Sect. 3 presents a description of the learning task, builds a rubric to evaluate activities, and summarises the progress of STEM activities. Then, the results are exhibited and analyzed in Sect. 4. Finally, we conclude the study and discuss future work in Sect. 5.

2 Related Work

The STEM education model has attracted widespread attention in the field of education. The combination of these fields to promote creative thinking, problem-solving, and the development of practical skills for students [7] has made significant marks in the education of research and implementation. The prospects of integrating STEM in education from early childhood to secondary level [8] provide an insight into the research that has been done and suggest a research agenda for educators' future STEM education. In this section, we present several specific studies related to the engineering design process in STEM, the results of which help optimally develop students' ability to solve real-world problems economically.

The authors in [9] proposed the application of STEM education by designing technical toys for middle school students. This study adopted a qualitative approach to execute integrated teaching within STEM education. The research was implemented in teaching technology subjects in Vietnamese middle schools. The design phase occurred at the Faculty of Technology Education of Hanoi National University of Education, Vietnam, in April 2015. The study addressed this challenge by exploring STEM education through experiential and creative activities in the Vietnamese educational landscape. Furthermore, it suggested a pragmatic model for integrating STEM into secondary school technology classes by creating technical toys. The interdisciplinary and cross-disciplinary integration methods align with the development of students' competencies outlined by the Vietnamese Ministry of Education. However, more incredible strides could be achieved by altering the curriculum to promote tighter integration across subjects, an avenue warranting further investigation. Another work in [10] addressed one approach to integrating the STEM disciplines within Grade 6 classes across three schools, where students engaged in engineering design processes as they created earthquake-resistant buildings. Considering several constraints, students employed engineering design processes and STEM disciplinary knowledge to plan, sketch, and construct a facility designed to withstand earthquake damage. In addition, the authors in [11] introduced the process of developing exercises for students using the STEM method to enhance their critical thinking skills. The study involved 36 11th-grade students from a public high school in Banjarmasin city as participants. The research instruments encompassed students' assignment sheets, lesson plan implementation forms, learning outcome assessment sheets, and an observation checklist for critical thinking skills. The results indicated that the student exercises using the STEM approach were suitable for cultivating necessary thinking skills within the learning environment. Moreover, assessing students' critical thinking skills was categorized as sufficient. In [12], the authors introduced the process of teaching a STEM topic in robotics to eighth graders—designing and manufacturing a vacuum cleaner robot. The theme is designed based on the engineering design process to create opportunities for students to develop problem-solving skills. Experimental results for 8th graders show that students' problem-solving ability is fostered, developed, and demonstrated during knowledge research, design, and manufacture of robot vacuum cleaners. In addition, the topic creates interest and curiosity in students in the field of robotics, and the process of organizing teaching is feasible in implementing and meeting the requirements of innovation in teaching content and teaching methods in schools today.

Studies have shown that implementing the engineering design process in STEM helps students develop essential life skills, including problem-solving, creative thinking, working, and problem-solving groups, and effectively apply theoretical knowledge in practice. In this paper, we present the technical process design of an automatic hand washing machine, from the selection of content to the method of instruction and evaluation.

3 Methods

In this section, we present a description of the learning task and build a rubric to evaluate the activities and progress of the STEM lesson on the design process of an automatic hand washing machine.

3.1 Description of Learning Tasks

A learning task is a task that requires students to apply knowledge and skills to solve a real-world problem and is assessed by clear criteria, detailing the requirements to be met.[1] In this study, we set out the task for students to design automatic hand washing machines to meet the following requirements:

Students designed an automatic hand washing machine to meet the following requirements:

- The product must operate automatically when the user puts his hand in the washing area.
- It needs to be designed to use the optimal amount of solution, which means calculating the amount of water required to wash hands effectively and ensuring no waste.
- Easy to use and spend less money to make the product.

[1] https://has.edu.vn/nhiem-vu-hoc-tap-giup-hoc-sinh-hinh-thanh-nang-luc-van-dung-kien-thuc-trong-thuc-tien/.

The two primary materials used to make the outer box of the product are cardboard paper and ABS plastic electrical boxes. Other materials include an infrared sensor, switch, power supply, pump, nozzle, and wire. To perform the task, students need to learn about the structure and operating principle of the relay switch, infrared sensor, and water pump mechanism. Students will apply their knowledge to answer the question, "How do we determine the amount of solution needed for the pump to be enough for one hand washing?".

3.2 *Building Rubric to Evaluate STEM Activities*

We built a rubric to evaluate the equality of members in each group, as shown in Table 1. With three levels, 0 is the lowest, and 2 is the highest. We care about cooperation, valuable contributions, contributions when making products, listening, and respecting everyone on the team.

Rubric, as shown in Table 2, evaluates technical design activities, including drawing circuit diagrams, designing product structure drawings, flowcharts of processing algorithms, expressions, and ways of answering questions that we care very carefully.

The product was made at home, so we asked the students to record a video. To evaluate the product's performance, we use a rubric as shown in Table 3 with three criteria for video evaluation, including the implementation content, the appearance of the video, and the washing machine product.

A rubric is shown in Table 4 for teachers to evaluate individual student activities during participation. Issues include learning sense, debate, exchange, feedback, cooperation, and time management.

Table 1 Assessment of equality in the group

TEAM MEMBERS' NAME	Cooperate			Valuable contribution in the group			Contributing during project implementation			Respect			Other notes (from other members)
	Enthusiastic, responsible, ready to accept work, ...			Answer, questions, give comments,...			Contributing efforts to the group's product implementation process			Listen, feedback, support,...			Things to promote, need to improve, Suggestions
	0	1	2	0	1	2	0	1	2	0	1	2	

Table 2 Rubric evaluates engineering design activity

Evaluation Criteria	Excellent	Good	Average	Need adjustment
	8–10 points	6–8 points	4–6 points	<4 points
Design drawing implementation	Scientific and rational. There are drawings of the expected product	Quite scientific and reasonable. There are expected product drawings but not detailed and clear	Not scientific or reasonable. There are expected product drawings but it is difficult to visualize	Unscientific, rational. No expected product drawings
Design electrical circuit diagrams	All materials needed are available. The layout diagram is clear, reasonable, and does not burn the circuit when in actual use	There are not enough materials to be used or the layout is not clear, and the circuit will not burn when in actual use	Insufficient materials needed. The layout diagram is not clear, there is a risk of circuit fire if used in practice	Insufficient materials needed. The circuit diagram has not been drawn
Design of processing algorithms	Draw algorithms to handle scientific and detailed problems	Draw algorithms to solve basic problems	Draw not the correct processing algorithm but understand the processing problem	Can't draw or draw wrong implementation principle
Presentation style	The presentation is engaging, the gestures and gestures are relevant to the content, covering the audience, and are interactive	Attractive presentation, gestures and gestures appropriate to the content, covering the audience but not interacting	Attractive presentation, gestures, gestures suitable for the content but not covering the audience	Gestures, gestures are not appropriate, have not covered the audience
Answers	Scientific explanation, coherent, convincing	Full explanation of the question, quite reasonable, quite convincing	The explanation is not convincing, the questions are not complete	The explanation is not convincing and not reasonable

Table 3 Rubric evaluates product design activities

Evaluation criteria	Excellent	Good	Average	Need adjustment
	8–10 points	6–8 points	4–6 points	<4 points
Video content	Fully demonstrate the process of making and finishing products. There are subtitles	Show quite fully the process of making and finishing products. There are subtitles	Inadequately show the process of making and perfecting the product. No subtitles available	The process of making and finishing products is not shown. No captions/subtitles
Video format	The picture is beautiful, clear, the sound is clear, the effects are suitable, there is creativity	The picture is beautiful, clear, the sound is clear, the effects are quite suitable, there's creativity	Images aren't beautiful, blurred, sound isn't clear, effects aren't suitable, lack of creativity	Images aren't beautiful, blurred, sound is not clear, abuse effects, are not creative
Hand washing machine products	The product has the ability to accurately recognize when someone puts their hands in and automatically activates the hand washing process. Beautiful products, ensuring safety for users, expressing creativity	The product has a relative ability to recognize when someone puts their hands in and automatically activates the hand washing process. The product is harmonious but not beautiful, ensuring safety for users	The product recognizes when someone places poor hands and automatically activates poor hand washing. The product isn't beautiful, not safe for users	The product doesn't recognize when someone puts their hands on it and doesn't trigger the hand washing process. The product hasn't been completed or isn't safe for users

4 Experiment

4.1 Experimental Object and Process

Experimental pedagogical research was conducted in 2 classes at An Khánh High School in Can Tho City, Vietnam. Each type is experimented with three lessons; each lesson is 45 min, design the curriculum framework as shown Fig. 1. According to the information provided on the ability to learn mathematics, physics, information technology, and technology from the subject teachers in charge, the students in the two classes have the same level, and the number of boys and girls in each category is approximately the same. Therefore, the experimental sample did not have a significant difference. Students work in groups with several 5–6 students per group, comprising eight groups.

Table 4 Rubric for teachers to evaluate individual students' activities during the participation process

Evaluation criteria	Excellent	Good	Average	Need adjustment
	8–10 points	6–8 points	4–6 points	<4 points
Consciously learning	Participate fully in classroom learning sessions and group work	Participate most of the time in classroom learning sessions and group work	Take the time to study in class and work in groups but let it go to waste	Participate but perform unrelated work
Debate, exchange, feedback	Pay attention to exchange, actively listen to the opinions of others and actively give personal opinions	Often listens carefully to the opinions of others, sometimes offering a personal opinion	Sometimes don't listen to the opinions of others. Rarely gives personal opinions or lacks initiative	Sometimes giving personal opinions but not listening to others
Cooperate	Respect the opinions of other members and cooperate with each other	Often respects the opinions of other members and cooperates in giving common opinions	Often respects the opinions of other members but hasn't yet cooperated to give a common opinion	Don't respect the opinions of other members and do not cooperate in giving common opinions
Time management	Complete assigned work on time	Often completes assigned work on time, without delaying the overall work of the team	Failure to complete assigned tasks on time and delay team work	Failure to complete assigned tasks on time and frequently forcing the team to adjust or change plans

4.1.1 Work 1: Problem Solution

Teachers let students watch videos about hand disinfection in hospitals during the COVID-19 outbreak. The teacher asked the students to fill in their feelings on the worksheet, which is the empathy chart. An empathy map is a visualization tool used during the design process to understand better a user's or target audience's emotions, opinions, needs, and desires. It helps create a holistic view of users by focusing on their emotional and experiential aspects. Empathy charts are often divided into sections like Fig. 2 to reflect elements of the user experience. During the activity, students summarized and suggested some ideas that need to be learned to solve the task, such as the solution automatically injected when the user puts their hand in.

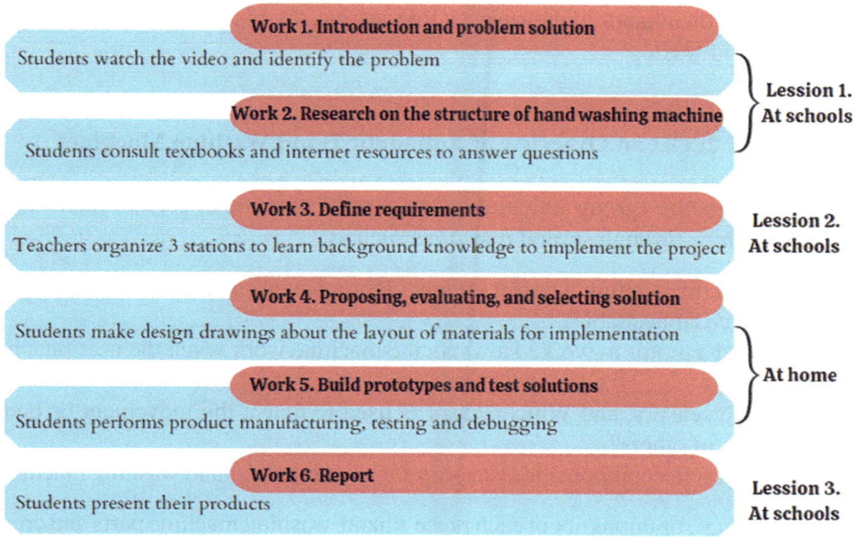

Fig. 1 The process of organizing experimental pedagogical teaching

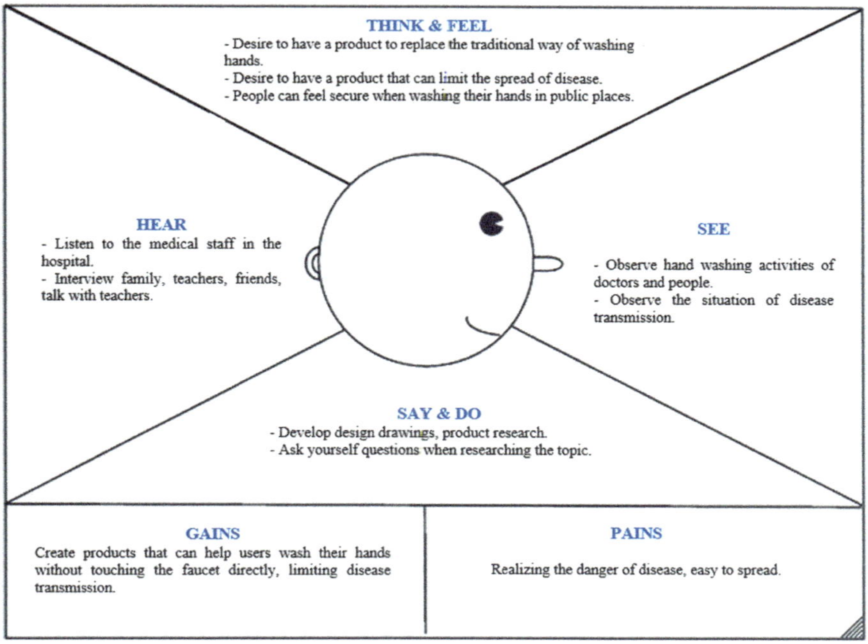

Fig. 2 Problem solving through empathy charts

When washing their hands, if they leave it for too long, the answer will not be pumped further to avoid waste.

4.1.2 Work 2: Get an Overview of Automatic Hand Washing Machines

The teacher organizes group activities on the general structure of the automatic hand washing machine through several questions such as:

- Does it work by needing a power supply? Use a battery or plug directly into the power source (if available).
- Does it need a motor to work? How can the machine work when the user puts his hand on it?
- Name some sensors, and Which sensor is used to detect the movement or body temperature of others?
- Please suggest the components necessary for an automatic hand washing machine.

Most groups have opinions about each piece's hand-washing machine parts and components. However, it needs to be clarified into three essential main parts: information processing unit sensors. Moreover, motives, an example image of a group learning about the structure, is shown in Fig. 3. For recommending components, the teams based on the hand washing machine model available in the market learn the principle and realize the task of pumping a disinfectant solution of the automatic hand washing machine. Observational results show that 6/8 groups (groups 1, 5, 8) can find infrared obstacle sensors, especially in group 5; there are also additional tools for washing machines. Automatic hand as a noncontact temperature sensor.

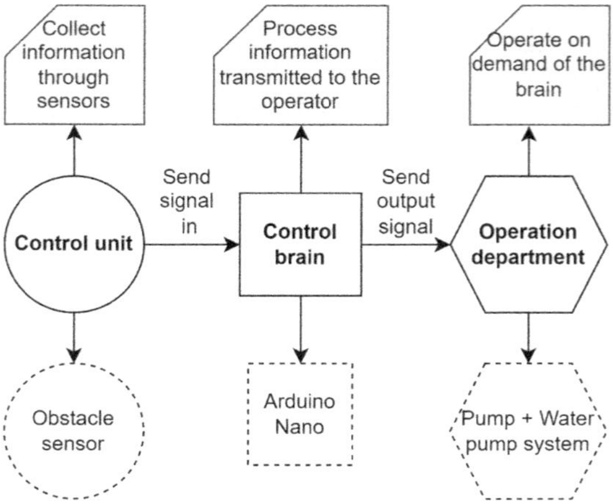

Fig. 3 Block diagram of the general structure of an automatic hand washing machine

4.1.3 Work 3: Define Requirements

The teacher organized three stations for this activity to study the background knowledge content to implement the product. Each student in the group will become a research expert for each station.

Station 1: Power source, motor, and current. Students watch a short video to study the knowledge and practice of drawing circuit diagrams and installing simple electrical circuits, including power sources, switches, and motors. At this station, almost 100% of the students could draw simple circuit diagrams, and 85% answered the correct theoretical questions about the current and power source.

Station 2: Signal, infrared sensor. Students study the practical knowledge of installing obstacle sensors and motors in Arduino for control. 80% of the children practiced correctly at this station. Because the sensor has many pins, some children could not distinguish it.

Station 3: Algorithm and programming block diagram. Students practice drawing block diagrams for the simple task "There is an obstacle in the pump motor". Teams analyze the task through the questions "How many input signals?", "How to activate the pump?", "How many seconds is enough for hand washing?" If the user gives "If the hand is continuously in the washing position, will it continue to pump the solution or stop?", "The Arduino's signal when it encounters an obstacle?" Most groups did an excellent job with this station. However, group 1 was unsuccessful in that if the user did not remove the hand, the solution would still pump. In group 1, students created more LED tools. If there is an obstacle except for starting the pump motor, they will light up the LED to distinguish that the machine is working. The processing algorithm of the automatic hand washing machine system is described in Fig. 4.

4.1.4 Work 4: Proposing, Evaluating, and Selecting Solutions

In this task, the teacher assigns the task to the students at home. Most groups can show automatic hand washing machine design drawings with complete parts such as materials, wiring diagrams, machine structural shapes, and algorithm flow diagrams for processing. However, groups still have difficulty because if users keep their hands fixed after washing the faucet, the machine will still pump the solution, causing waste. At that time, there was a group that suggested that when programming the product, the condition would be used in addition to detecting obstacles to turning on the pump; if there are no obstacles to turning off the pump, they will add time, on average each person washing their hands will It takes 2 s to pump the solution, if the obstruction remains for more than 2 s, the pump will turn off instead of pumping continuously. Figure 5 shows the wiring diagram of the automatic washing machine system.

4.1.5 Work 5: Build Prototypes and Test Solutions

In this activity, students make products, test, and fix errors in products at home. During implementation, students use cardboard, forex, and plastic boxes to process

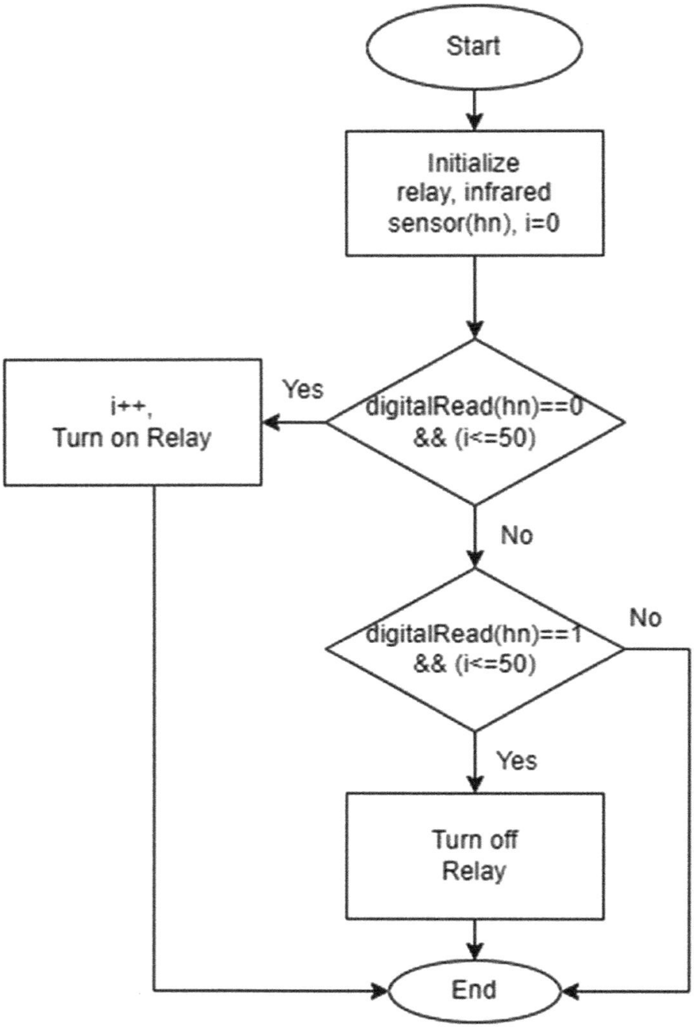

Fig. 4 Algorithm flowchart of a group of students

hand washing machines. The process of designing hand-washing pump parts for each group has different ideas. However, they all share the same purpose of making an automatic solution pumping system and not consuming too much solution. However, some groups design the solution container to the outside of the machine to easily add washing solution when used; some groups are left inside to increase the machine's aesthetics. Most teams can solve programming tasks when there is an obstacle, and the pump will work. However, many groups have difficulty solving the problem of immobilizing obstacles, but the pump still works. The additional group of 8 has obstacles; then turn on the LED to recognize that the machine is operating.

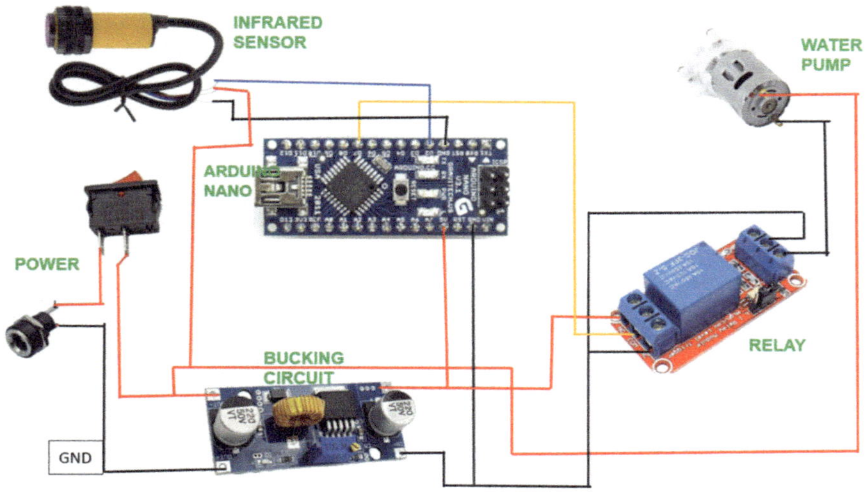

Fig. 5 Results of wiring diagram reasoning of a group of students

4.1.6 Work 6: Report

In the last activity, students share about the design and finished product. Students can generally present confidently, showing two important contents as required: the argument for calculating the number of materials and the installation layout. During the design process, the students became aware that a sensor system was required for an automatic hand-washing machine to work effectively. This allows the machine to detect the user's presence and automatically trigger the hand-washing process. At the same time, it is essential to control the amount of hand sanitizer needed. They suggested adjusting the time to wash hands by no more than 5 s each time to ensure that the solution is enough to remove bacteria without causing waste. In addition, they also calculate the needs of users. If the washing machine is used in a practical environment and the number of users is large, the solution tank should be located outside for easy replenishment when it runs out. This ensures the design's aesthetics, convenience, and efficiency in use. Figure 6 is one of the group's report posters.

4.2 Evaluation

4.2.1 Design

Students have designed many versions of automatic hand sanitizer products, including slurry and alcohol solutions. Depending on specific usage needs, they will create solution containers inside and outside the product during the design process. For example, if the product is placed in a classroom, the solution container will be inside

Fig. 6 Poster of a group of students reporting the product

to tidy the space. On the contrary, if the product is located in a school lobby with many users, the solution tank will be placed outside to quickly replace and replenish the solution.

Regarding the design of the washing nozzle, depending on usage needs, it can be adjusted to suit each type of solution. This helps avoid clogged faucets or wasting solutions. By adjusting the cleaning nozzle, the product can effectively use slurry and alcohol solutions without causing any problems. This demonstrates the students' creativity and meticulous consideration in designing the automatic hand-washing machine product.

Students designed the product, so its form is less aesthetically pleasing than existing industrial products. However, the product still ensures functionality and is used effectively, demonstrating innovation in students' creativity and efforts in applying STEM knowledge into practice. The most important thing is that the product is capable of performing its task, an automatic hand washing machine that can effectively clean hands and save liquid, and at the same time save many costs for the school, instead of having to If we spend 80 USD to buy one automatic hand washing machine, we can make six others with the same function. The children designed some automatic hand washing machines as shown in Fig. 7.

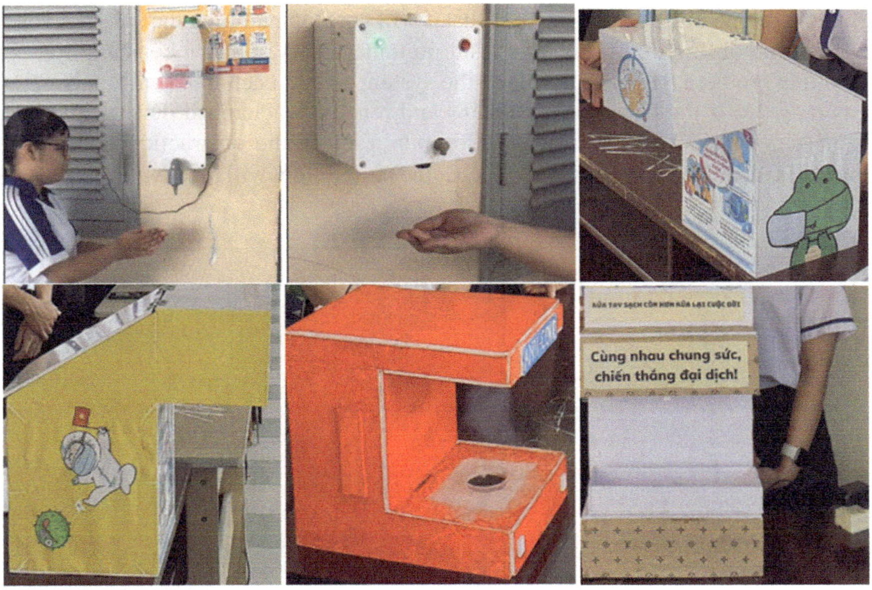

Fig. 7 Some finished products designed by students and put into use in schools

4.2.2 Save on Hand Sanitizer

According to the recommendations of the Ministry of Health of Vietnam, each time we wash with an alcohol solution, we will need about 3–5ml and wash our hands for 15 to 20 s. However, when the actual use (not yet applied automatic hand washing machine), according to our survey at An Khanh High School at the time after the Covid-19 epidemic, each 1-liter bottle of solution the children used about 146 to 173 students (each student corresponds to 1 wash when entering class).

In this study, We used a mist sprayer with an operating pressure of 2.5Pa and a water flow rate ranging from 9 to 12 liters per hour when using an alcohol solution. This means that every second, the nozzle can deliver at least 2.5ml of solution (depending on the adjustment via the screwdriver inside the nozzle). Students have made calculations to ensure enough solution for each hand wash, namely 5ml; the nozzle needs to work for about 2 s to ensure enough solution for the hand washing process. From there, performing about 200 hand washes with 1 liter of solution is possible. Then, we experimented with students by designing a solution container with a capacity of 1 liter to be placed outside to monitor the amount of solution used. This experiment was conducted for 30 consecutive days in September 2021, and we collected detailed data as depicted in Fig. 8.

One possible situation is when a person washes their hands for too long; the principle of operation of the sensor is that it will not stop pumping the solution when it detects an obstruction. This may result in unnecessary solution consumption. To solve this problem, we have designed the system so that each handwashing session lasts only 2 s. If the obstruction remains and remains unchanged during this time, the system will not activate the solution pump. Once the obstruction is removed from the sensor position and reintroduced, the new system will reactivate the process

Fig. 8 Product performance for 30 consecutive days with 1 liter of alcohol solution

as usual. This helps to avoid unnecessary waste of solution and ensures efficiency during automatic handwashing.

4.3 Discussion

In general, in the process of participating in a STEM lesson on the topic of automatic hand washing machines, the majority of student groups have shown positive signs in learning, discussing, and proposing solutions for the design. They showed passion and interest in studying science and techniques' scientific and technical principles. Groups of students came together to learn about the structure, working principles, and other essential elements of automatic hand-washing machines. They discussed and exchanged ideas and showed creativity in developing design solutions based on the collected knowledge. They have learned from each other, sharing knowledge and experiences to achieve common goals. The groups evaluated and improved their solutions through discussion and exchanging ideas, creating a dynamic and creative learning process. For example, two groups of students showed interest in the topic when they made a finished product model at home and put it to practical use in school.

Besides, some difficulties are also encountered in the process of product implementation. The first thing that can be mentioned is the need for significant investment capital for equipment. Components such as sensors, pumps, control systems, and high-quality materials may not be easily accessible to groups of students. That can limit the ability to implement complete design solutions and make it difficult to perform tests and checks. Second, it requires deep knowledge of electronics and programming. Our target audience is high school students at the age of 16, so mechanical engineering to make products as beautiful as designs is a matter of concern; in addition, they have yet to learn deeply about engineering programmed to handle intensive problems in real-world testing.

5 Conclusion

Automated hand washing machine STEM topic is an integral part of Vietnam's 2018 general education program, aiming to foster and develop students' problem-solving abilities in Natural Sciences. Nature, Technology, Informatics and Mathematics. Experimental results have shown that the automatic hand washing machine STEM topic has significantly helped students develop problem-solving abilities. Students could apply knowledge from various STEM subjects and combine research, analysis, creativity, and logical thinking skills to design and build an automated hand-washing machine model. In addition, students develop the ability to work in groups, communicate, present ideas, and learn from each other. They experienced learning, experimentation, and improvement, discovering how to solve problems and discovering their potential.

References

1. Hsu, Y.S., Fang, S.C.: Opportunities and challenges of STEM education. In: Asia-Pacific STEM Teaching Practices, pp. 1–16. Springer Singapore (2019). https://doi.org/10.1007/978-981-15-0768-7_1
2. Holstermann, N., Grube, D., Bögeholz, S.: Hands-on activities and their influence on students' interest. Res. Sci. Educ. **40**(5), 743–757 (2009). https://doi.org/10.1007/s11165-009-9142-0
3. Shahali, E.H.M., Halim, L., Rasul, M.S., Osman, K., Zulkifeli, M.A.: STEM learning through engineering design: impact on middle secondary students' interest towards STEM. EURASIA J. Math., Sci. Technol. Educ. **13**(5) (2016). https://doi.org/10.12973/eurasia.2017.00667a
4. Owston, R., York, D.N., Malhotra, T., Sitthiworachart, J.: Blended learning in STEM and non-STEM courses: how do student performance and perceptions compare? Online Learn. **24**(3) (2020). https://doi.org/10.24059/olj.v24i3.2151
5. Hiğde, E., Aktamış, H.: The effects of STEM activities on students' STEM career interests, motivation, science process skills, science achievement and views. Think. Ski. Creat. **43**, 101000 (2022). https://doi.org/10.1016/j.tsc.2022.101000
6. Stevens, D.D., Levi, A.J.: Introduction to rubrics: an assessment tool to save grading time, convey effective feedback, and promote student learning. Routledge (2023)
7. Council, N.R., et al.: Successful K-12 STEM education: identifying effective approaches in science, technology, engineering, and mathematics. National Academies Press (2011)
8. Council, N.R., et al.: STEM integration in K-12 education: status, prospects, and an agenda for research. National Academies Press (2014)
9. Quang, L.X., Hoang, L.H., Chuan, V.D., Nam, N.H., Anh, N.T.T., Nhung, V.T.H.: Integrated science, technology, engineering and mathematics (stem) education through active experience of designing technical toys in vietnamese schools (2015). arXiv:1509.03807
10. English, L.D., King, D., Smeed, J.: Advancing integrated STEM learning through engineering design: sixth-grade students' design and construction of earthquake resistant buildings. J. Educ. Res.**110**(3), 255–271 (2016). https://doi.org/10.1080/00220671.2016.1264053
11. Hartini, S., Mariani, I., Misbah, Sulaeman, N.F.: Developing of students worksheets through STEM approach to train critical thinking skills. J. Phys.: Conf. Ser. **1567**(4), 042029 (2020). https://doi.org/10.1088/1742-6596/1567/4/042029
12. Phuong, D.D., Thang, V.Q., Anh, N.D., Ngan, L.H.M.: Stem education: simple robot vacuum cleaner based on the engineering design process for secondary students. J. Sci. **18**(8), 1495 (2021). https://doi.org/10.54607/hcmue.js.18.8.3022(2021)

KNN-SMOTE: An Innovative Resampling Technique Enhancing the Efficacy of Imbalanced Biomedical Classification

Xuan Tho Dang and Thanh Tuoi Le

Abstract In recent years, the prevalence of imbalanced data has escalated within real-world problem domains, inducing substantial challenges that impede data analysis and exploration. This predicament is particularly pronounced within biomedical data classification tasks, wherein the imbalance quandary has garnered notable attention from the data science and machine learning research community. Various approaches, notably SMOTE and its variants, have been developed to address this concern and have demonstrated encouraging outcomes. Nonetheless, a comprehensive resolution to this issue remains elusive, with instances where the effectiveness of these methods falters or even diminishes classification performance. Consequently, this paper introduces a novel method named KNN-SMOTE, an enhancement of the Borderline-SMOTE technique. Empirical assessments conducted on five imbalanced benchmark datasets from the UCI Machine Learning Repository underscore the superior performance of our approach in terms of F-score, G-mean, and AUC in comparison to other methodologies.

Keywords SMOTE · Safe-level SMOTE · Borderline-SMOTE · Imbalanced dataset · KNN-SMOTE

1 Introduction

In the present era, as the significance of big data continues to grow, research focused on the processing and exploration of large datasets has become a prominent field. This poses challenges to conventional machine learning methods, which aim for speed, efficiency, and precision. Currently, there is a lack of efficient approaches to

X. T. Dang
Academy of Policy and Development, Ha Noi, Viet Nam
e-mail: thodx@apd.edu.vn

T. T. Le (✉)
Vinh University of Technology and Education, Vinh, Viet Nam
e-mail: fredleektv@gmail.com

N. Hoang Phuong et al. (eds.), *Machine Learning and Other Soft Computing Techniques: Biomedical and Related Applications*, Studies in Systems, Decision and Control 543, https://doi.org/10.1007/978-3-031-63929-6_11

mine various real-world data types, including tasks such as identifying fraudulent credit card transactions [1], detecting unauthorized network intrusions [2], spotting oil spills from satellite images [3], and making diagnoses and predictions in the field of biomedicine [4].

Notably, a major challenge faced by data scientists is the increasing prevalence of imbalanced datasets. Conventional machine learning algorithms are built on the assumption of a balanced distribution of samples across various classes. However, the presence of imbalanced sample sizes poses challenges for algorithms biased towards the majority class. Despite the reduced number of samples, the minority classes often hold higher significance. Therefore, there is a need to enhance the accuracy of learning algorithms in correctly diagnosing cases within these minority classes [5]. The misclassification of a minority sample can carry severe consequences, especially in fields like disease diagnosis and patient identification. It is imperative that an optimal classification model prioritizes achieving a higher rate of accurate diagnosis for diseases within the minority class [6].

An instance showcasing the real-world implications of imbalanced learning can be found in the biomedical field, specifically through a dataset gathered from a series of mammography X-ray images of different patients. This scenario underscores the concept of imbalanced learning. The binary image analysis establishes classes (labels) where patients are either classified as normal or having cancer, corresponding to positive or negative labels. Based on experience, it's expected that the number of patients without cancer (negative class) will far outnumber those with cancer (positive class). The dataset comprises 10,923 negative samples (majority class) and 260 positive samples (minority class). An optimal classifier should maintain a balanced ratio for predicting both the majority and minority classes in the dataset with equal accuracy (ideally 100%). However, real-world classification demonstrates a high degree of imbalance, with 100% accuracy for the majority group and 0% to 10% for the minority group. With a 10% accuracy rate for classifying the minority group, this implies that 234 patients within this group will be inaccurately classified as part of the majority. Consequently, 234 cancer patients will be wrongly diagnosed as non-cancerous. In the medical field, errors of this nature incur a higher cost than misclassifying a non-cancer patient as having cancer [7].

Hence, it holds significant importance to engage in research concerning the issue of class imbalance. The challenge of class imbalance is widely prevalent and encompasses numerous crucial domains in data analysis. Additionally, the majority of machine learning classification systems have been reported to lack the capacity to effectively address the problem of class imbalance. In this article, we illustrate that in certain cases, many methods among these become unsuitable and fail to generate useful artificial samples for the minority class. We also highlight situations in which the methods would produce erroneous and unnecessary synthetic samples, thereby complicating the learning process. In this regard, we introduce a novel oversampling method, namely the K-Nearest Neighbors Synthetic Minority Over-sampling Technique (KNN-SMOTE), aimed at mitigating imbalanced learning issues and producing meaningful artificial samples for the minority class. The main contributions of this study are outlined as follows:

(a) We present and analyze the strengths and weaknesses of several prior data balancing methods, leading to the proposal of an enhanced approach.
(b) In response to the challenge of data imbalance, we introduce a novel enhancement method that effectively mitigates this issue by leveraging nearest neighbor elements.
(c) Through experiments conducted on biomedical datasets, we demonstrate that the application of these methods improves classification performance compared to previous techniques.

The subsequent sections of this paper are organized as follows: Sect. 2 offers a comprehensive survey of existing literature in the domain, assessing the merits and drawbacks of earlier methodologies. Our proposed approach is delineated in Sect. 3, elucidating its fundamental constituents and strategies. Section 4 details the experimental results, their analysis shedding light on the efficacy and performance of our approach. Finally, Sect. 5 encapsulates our findings, draws conclusions based on the results, and deliberates on potential directions for future research and advancement.

2 Related Work

Due to the significance of imbalanced datasets, research addressing the class imbalance issue remains a continual endeavor. The objective of such investigations is to optimize accurate diagnosis within the minority class. Different avenues have yielded distinct methodologies designed to tackle this challenge, which can be broadly categorized into two primary groups:

(a) Data-Level Solutions: The first approach involves preprocessing the input data to attain class balance by redistributing instances across the data space. This equilibrium can be achieved by either augmenting the minority class or reducing the majority class. These methods are commonly referred to as data preprocessing techniques [8, 9].
(b) Algorithmic-Level Solutions: In this category, the learning algorithm itself is either modified or tailored to effectively handle imbalanced datasets [10–12].

This article focuses on the second category, the Data-Level Solutions. Methods falling under this group alter the distribution of the minority and majority classes. The objective of this process, known as resampling, aims to rebalance or restore a reasonable equilibrium achieved by either removing instances of the majority class or introducing instances of the minority class. Two techniques for this purpose are referred to as undersampling and oversampling, and they can be combined as a hybrid method [12–14]. Resampling is further categorized into three groups:

(i) Oversampling: This involves identifying and enhancing the minority class by generating new samples or replicating existing ones.
(ii) Undersampling: This equalizes samples from both classes by reducing data and eliminating instances from the majority class.

(iii) Hybrid method: This approach combines the two resampling techniques. The utilization of synthetic oversampling techniques has proven highly effective in addressing imbalanced data scenarios [15–22]. Nonetheless, our investigation reveals certain deficiencies and inadequacies in the current methods, which could manifest across various scenarios.

For instance, in the case of the Borderline-SMOTE [16] method, the primary objective is to identify minority class samples along the border. The oversampling technique utilizes these borderline instances to generate synthetic samples, as they are prone to misclassification by the classifier. This is why Borderline-SMOTE creates synthetic samples in the vicinity of the border. According to, a minority class sample is designated as a border sample if the number of majority class samples, denoted as m, among its k nearest neighbors satisfies the condition:

$$\frac{k}{2} \leq m < k \tag{1}$$

where k is a user-specified parameter. However, this criterion may fail to identify border instances in some cases. Figure 1 depicts such a scenario where the minority class instances x and y do not have any majority class instances in their k nearest neighbors' vicinity (assuming $k = 5$). It can be observed from Fig. 1 that the 5 nearest neighbors of x (indicated by arrows) are minority class samples. Thus, both x and y will have $m = 0$. According to equation (1), i.e., $m = 0$ and $k = 5$, the opportunity for x and y to be chosen as border instances will be eliminated, even though they are the nearest samples to the border. Consequently, no synthetic instances will be generated from x and y. This situation can complicate the learning task as there is insufficient information about the minority class instances near the border.

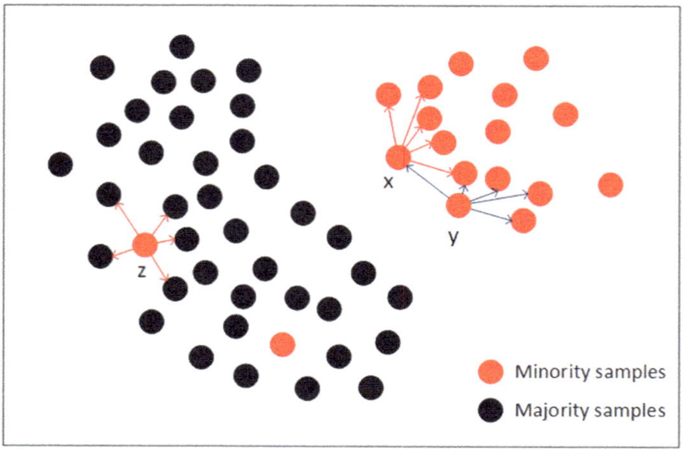

Fig. 1 The limitations of the Borderline-SMOTE method, with the k-nearest neighbors of x, y and z shown using an arrow ($k = 5$)

3 The Method

Motivated by the challenges highlighted in the preceding section, we introduce a novel method for minority oversampling, namely KNN-SMOTE. The goals of this method encompass two main aspects: enhancing the process of selecting borderline minority instances and refining the generation of synthetic samples. Our KNN-SMOTE methodology comprises four primary stages. In the initial phase, KNN-SMOTE constructs a filter to eliminate instances of the minority class lacking any neighboring minority instances, resulting in a filtered minority dataset denoted as S_{minf}. In the second phase, based on S_{minf}, we aggregate a set of enhanced versions of the majority class instances that contain significant information, referred to as S_{bmaj}. Subsequently, in the third phase, KNN-SMOTE identifies a collection of instances from the minority class residing in the vicinity of important class boundaries, termed as S_{imin}. Finally, the synthetic versions are synthesized by applying the SMOTE algorithm to the S_{imin} dataset. The entire process is elucidated in [Algorithm 1]. Steps 1 and 2 represent the first phase, steps 3 and 4 embody the second phase, steps 5 and 6 illustrate the third phase, and lastly, step 7 outlines the termination phase. The key components of KNN-SMOTE are described as follows:

In Fig. 2a, for each element $x_i \in S_{min}$, calculate the nearest neighbors set $NN(x_i)$. $NN(x_i)$ consists of the $k1$ nearest neighbors of element x_i based on the Euclidean distance. Construct the under-sampled set S_{minf} from the S_{min} set by removing the elements from the minority class whose neighbors do not belong to the minority

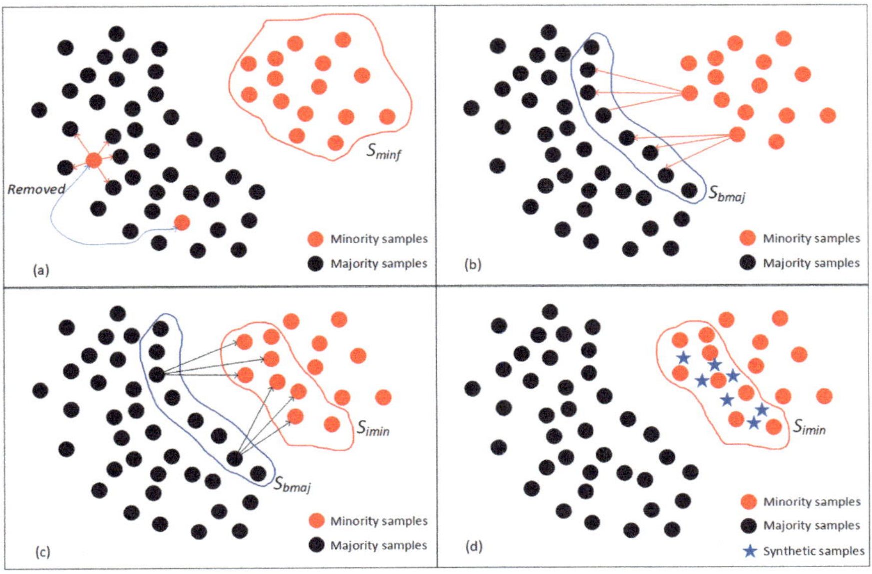

Fig. 2 The execution steps of the novel proposed method, KNN-SMOTE

Algorithm 1: KNN-SMOTE(S_{maj}, S_{min}, N, $k1$, $k2$, $k3$)

Input:
S_{maj}: Set of majority class instances
S_{min}: Set of minority class instances
N: Number of additional synthetic instances to be generated
$k1$: Number of neighbors used to filter noisy minority class samples
$k2$: Number of majority class neighbors employed to construct a set of significant information-bearing majority class elements.
$k3$: Number of minority class neighbors used to build a collection of important-information-containing minority class elements.

Procedure Commencement:
Step 1. For each element $x_i \in S_{min}$, calculate the set of $k1$ nearest neighbors $NN(x_i)$. $NN(x_i)$ includes the $k1$ closest neighbors of x_i based on Euclidean distance.
Step 2. Construct the filtered minority set S_{minf} by removing minority class instances whose neighbors do not contain any minority class instances.
$S_{minf} = S_{min} - \{x_i \in S_{min} : NN(x_i) \text{ does not contain any minority class instances}\}$
Step 3. For each $x_i \in S_{minf}$, compute the set of nearest majority instances $N_{maj}(x_i)$. $N_{maj}(x_i)$ consists of the $k2$ nearest majority class neighbors of x_i based on Euclidean distance.
Step 4. Determine the majority class borderline set S_{bmaj} by uniting all $N_{maj}(x_i)$.
$S_{bmaj} = \bigcup_{x_i \in S_{minf}} (N_{maj}(x_i))$
Step 5. For each $y_i \in S_{bmaj}$, calculate the set of nearest minority instances $N_{min}(y_i)$. $N_{min}(y_i)$ comprises the $k3$ nearest minority class neighbors of y_i based on Euclidean distance.
Step 6. Find the minority class borderline set S_{imin} by uniting all $N_{min}(y_i)$.
$S_{imin} = \bigcup_{y_i \in S_{bmaj}} (N_{min}(y_i))$
Step 7. Apply the SMOTE algorithm on the dataset S_{imin} to generate additional synthetic minority class instances.

class. $S_{minf} = S_min - \{x_i \in S_min : NN(x_i)$: doesn't contain minority class elements$\}$ (steps 1–2 of the KNN-SMOTE algorithm).

Figure 2b illustrates for each $xi \in Sminf$, create the set of majority class elements $N_{maj}(x_i)$. This is done so that the samples in $N_{maj}(x_i)$ will be the majority class elements containing important information and are expected to be located near the class boundary when $k2$ is small. Combine all $N_{maj}(x_i)$ to form the set of majority class elements with important information, S_{bmaj} (steps 3–4 of the KNN-SMOTE algorithm).

Next, in Fig. 2c, for each element $y_i \in S_{bmaj}$, construct the set of nearest minority class elements with important information, $N_{min}(y_i)$. $N_{min}(y_i)$ is the collection of $k3$ nearest minority class elements to y_i based on Euclidean distance. Combine all such $N_{min}(y_i)$ to create S_{imin} (steps 5–6 of the KNN-SMOTE algorithm).

Lastly, apply the SMOTE algorithm to generate additional synthetic instances within the S_{imin} dataset as shown in Fig. 2d.

4 Experiments

4.1 Datasets

In order to assess the effectiveness of the novel algorithm KNN-SMOTE, the authors conducted experiments on five imbalanced datasets sourced from the internationally recognized UCI dataset repository [6]. The UCI dataset repository encompasses various datasets from diverse domains. Within the scope of this study, the experiments were conducted on the following specific datasets (Table 1).

All the datasets exhibit class imbalance with two class labels: Positive (minority class) and Negative (majority class). The imbalance ratios vary across the datasets: Yeast dataset has a ratio of 1:28, Blood, Abalone, and Coil2000 have ratios of 1:16, and Ionosphere has ratios of 1:2. Among these, the dataset with the largest number of instances is Coil2000 with 5,822 elements and 86 features.

4.2 Evaluation Measures

For computing metrics assessing classification performance, we employ a confusion matrix outlined in Table 2. In this matrix, the minority class instances are denoted as Positive, while the majority class instances are marked as Negative.

For each dataset, following the classification process, we calculate the counts of TP, FP, TN, and FN in a statistical manner. These values are then utilized to compute metrics aimed at assessing the efficiency of classification. Certain metrics are derived from the information contained in the confusion matrix:

Table 1 Illustrates five datasets extracted from the internationally recognized UCI dataset repository

Data	Samples	Features	Imbalance ratios (Positive/Negative)
Yeast	1484	8	1:28
Blood	731	8	1:16
Abalone	731	8	1:16
Coil2000	5822	86	1:16
Ionosphere	351	35	1:2

Table 2 Confusion matrix

	Predicted as positive	Predicted as negative
Reality positive	True Positives (TP)	False Negatives (FN)
Reality negative	False Positives (FP)	True Negatives (TN)

$$TPR = recall = sensitivity = \frac{TP}{TP + FN} \tag{2}$$

$$TNR = specificity = \frac{TN}{TN + FP} \tag{3}$$

$$PPV = precition = \frac{TP}{TP + FP} \tag{4}$$

$$ACC = \frac{TP + TN}{TP + FP + TN + FN} \tag{5}$$

Accuracy serves as a metric to assess classification performance in balanced datasets. Nonetheless, this metric doesn't effectively gauge the performance of imbalanced datasets. This is due to the overwhelming number of majority class samples, leading to a high accuracy even when the minority class samples are predicted at a significantly low percentage. Therefore, a value that reflects the equilibrium between *TPR*, *TNR*, and *PPV* is computed through a numerical formula [12, 13]:

$$F1 - score = \frac{2 * PPV * TPR}{PPV + TPR} \tag{6}$$

$$G - mean = \sqrt{TPR * TNR}. \tag{7}$$

4.3 *Classification Imbalance Learning Results*

To assess the effectiveness of the methods, we conducted experiments in R using five datasets. Subsequently, we compared the classification performance of the adjusted datasets using the proposed method (KNN-SMOTE) with popular methods such as SMOTE and its variations, including Borderline-SMOTE [17], Safe-level SMOTE [18], SMOTE-PSO [21], and SMOTE-Kmean [22]. In this study, we employed the Support Vector Machines (SVM) classification algorithm using the Kernlab package [23] with default parameters to classify the original datasets and the balanced ones. To ensure objective evaluation and minimize overfitting, we conducted tests using the 10-fold cross-validation technique.

The F-score, G-mean, and AUC values are employed for comparative analysis across each machine learning method. Figure 3 illustrates the comparison of training and testing outcomes using different data balancing techniques based on the F-score metric. By examining the table, it's evident that the KNN-SMOTE method yields higher G-mean values, surpassing most other methods by 1–3% compared to the scenario without KNN-SMOTE.

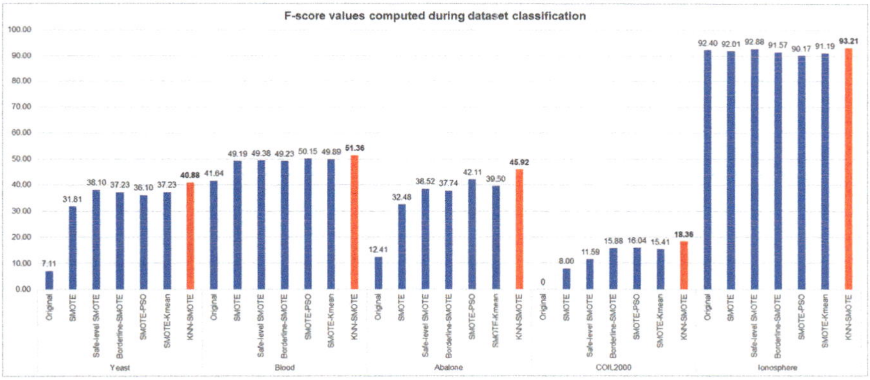

Fig. 3 F-score values computed during dataset classification

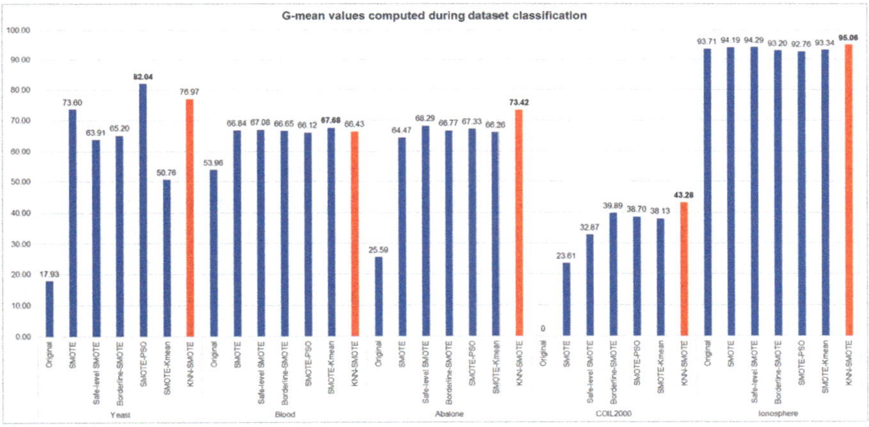

Fig. 4 G-mean values computed during dataset classification

Furthermore, the G-mean metric, which acts as a classifier's measure balancing Sensitivity and Specificity, is also employed to verify the accuracy of prediction models (Fig. 4). The results presented in Fig. 4 demonstrate that the G-mean values of models utilizing KNN-SMOTE are higher than those of models trained on the original datasets and the SMOTE, Safe-level SMOTE, Borderline-SMOTE, SMOTE-PSO, and SMOTE-Kmean methods in most cases of imbalanced data.

Moreover, the experimental results and AUC-based evaluation also revealed that our method achieves significantly better AUC values compared to the control method, SMOTE, and its variations as shown in Fig. 5.

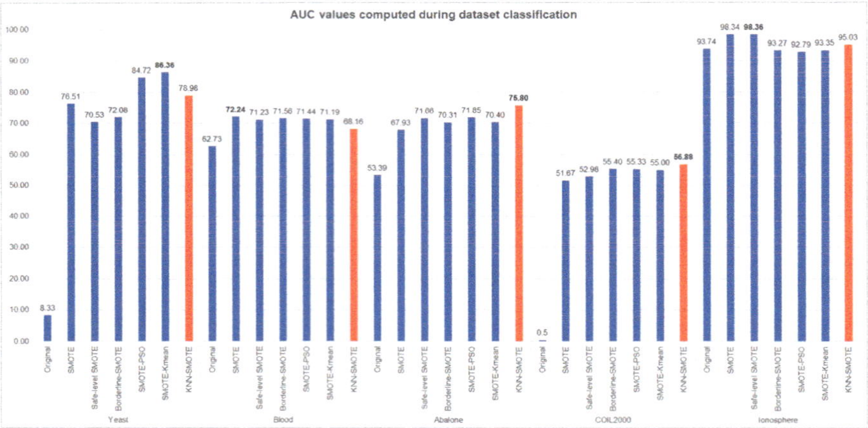

Fig. 5 AUC values computed during dataset classification

5 Conclusions

In this paper, we have presented and analyzed the strengths and weaknesses of several previous data balancing methods such as SMOTE and its variants. Building upon the insights gained from these analyses, we introduce an advanced approach aimed at effectively mitigating these issues by harnessing the nearest neighbor factors. Experimental results conducted on biomedical datasets have demonstrated that the application of these methodologies contributes to an enhanced classification performance compared to prior techniques. We firmly believe that the realm of this research will continue to offer abundant avenues for scholars to delve further, investigate, and refine, thus further elevating the efficacy in addressing the challenging imbalance data problem.

References

1. Ahmed, M., Mahmood, A.N., Islam, M.R.: A survey of anomaly detection techniques in financial domain. Futur. Gener. Comput. Syst. **55**, 278–288 (2016)
2. Zareapoor, M., Shamsolmoali, P.: Application of credit card fraud detection: based on bagging ensemble classifier. Procedia Comput. Sci. **48**(2015), 679–685 (2015)
3. Chen, G., Li, Y., Sun, G., Zhang, Y.: Application of deep networks to oil spill detection using polarimetric synthetic aperture radar images. Appl. Sci. **7**(10), 968 (2017)
4. Jia, J., Liu, Z., Xiao, X., Liu, B., Chou, K.C.: Ippbs-Opt: a sequence-based ensemble classifier for identifying protein-protein binding sites by optimizing imbalanced training datasets. Molecules **21**(1), 95 (2016)
5. Krawczyk, B.: Learning from imbalanced data: open challenges and future directions. Prog. Artif. Intell. **5**(4), 221–232 (2016)

6. Sun, Y., Wong, A.K., Kamel, M.S.: Classification of imbalanced data: a review. Int. J. Pattern Recognit. Artif. Intell. **23**(04), 687–719 (2009)
7. He, H., Garcia, E.A.: Learning from imbalanced data. IEEE Trans. Knowl. Data Eng. **21**(9), 1263–1284 (2009)
8. Longadge, R., Dongre, S.: Class imbalance problem in data mining review (2013). Arxiv:1305.1707
9. Hung, B.D., Anh, D.N., Tho, D.X.: Relabeling with Mask-S for imbalanced class distribution. In: Frontiers In Intelligent Computing: Theory and Applications: Proceedings of the 7th International Conference on FICTA (2018), vol. 1, pp. 31–41. Springer Singapore (2020)
10. Majid, A., Ali, S., Iqbal, M., Kausar, N.: Prediction of human breast and colon cancers from imbalanced data using nearest neighbor and support vector machines. Comput. Methods Programs Biomed. **113**(3), 792–808 (2014)
11. Devi, D., Purkayastha, B.: Redundancy-driven modified Tomek-link based undersampling: a solution to class imbalance. Pattern Recogn. Lett. **93**, 3–12 (2017)
12. Anh, D.N., Hung, B.D., Huy, P.Q., Tho, D.X.: Feature analysis for imbalanced learning. J. Adv. Comput. Intell. Intell. Inform. **24**(5), 648–655 (2020)
13. Dang, X.T., Bui, D.H., Nguyen, T.H., Nguyen, T.Q.V., Tran, D.H.: Prediction of autism-related genes using a new clustering-based under-sampling method. In: 2019 11th International Conference on Knowledge and Systems Engineering (KSE), pp. 1–6. IEEE (2019)
14. Dang, X.T., Hirose, O., Bui, D.H., Saethang, T., Tran, V.A., Nguyen, L.A.T., Satou, K.: A novel over-sampling method and its application to cancer classification from gene expression data. Chem-Bio Inform. J. **13**, 19–29 (2013)
15. Sáez, J.A., Luengo, J., Stefanowski, J., Herrera, F.: SMOTE-IPF: Addressing the noisy and borderline examples problem in imbalanced classification by a re-sampling method with filtering. Inf. Sci. **291**, 184–203 (2015)
16. Chawla, N.V., Bowyer, K.W., Hall, L.O., Kegelmeyer, W.P.: SMOTE: synthetic minority oversampling technique. J. Artif. Intell. Res. **16**, 321–357 (2002)
17. Han, H., Wang, W.Y., Mao, B.H.: Borderline-SMOTE: a new over-sampling method in imbalanced data sets learning. In: International Conference on Intelligent Computing, pp. 878–887. Springer Berlin Heidelberg, Berlin, Heidelberg (2005)
18. Bunkhumpornpat, C., Sinapiromsaran, K., Lursinsap, C.: Safe-Level SMOTE: safe-level-synthetic minority over-sampling technique for handling the class imbalanced problem. In: Advances in Knowledge Discovery and Data Mining: 13th Pacific-Asia Conference, PAKDD 2009 Bangkok, Thailand, April 27–30, 2009 Proceedings 13, pp. 475–482. Springer Berlin Heidelberg (2009)
19. Tho, D.X., Anh, D.N.: Imbalance In Learning Chest X-Ray images For COVID-19 detection. In: Soft Computing: Biomedical and Related Applications, pp. 107–119. Springer International Publishing, Cham (2021)
20. Dang, X.T., Tran, D.H., Hirose, O., Satou, K.: SPY: a novel resampling method for improving classification performance in imbalanced data. In: 2015 Seventh International Conference on Knowledge and Systems Engineering (KSE), pp. 280–285. IEEE (2015)
21. Cervantes, J., Garcia-Lamont, F., Rodriguez, L., López, A., Castilla, J.R., Trueba, A.: PSO-based method for SVM classification on skewed data sets. Neurocomputing **228**, 187–197 (2017)
22. Douzas, G., Bacao, F., Last, F.: Improving imbalanced learning through a heuristic oversampling method based on K-means and SMOTE. Inf. Sci. **465**, 1–20 (2018)
23. Karatzoglou, A., Smola, A., Hornik, K.: kernlab: kernel-based machine learning lab. R Package Version 0.9–32 (2023)

Human Detection in Video for Security Surveillance Systems

Nguyen Thai-Nghe, Pham Thanh-Son, and Nguyen Huu-Hoa

Abstract Human detection in video is an important task for security surveillance systems. It could help reducing time and effort when tracking a human in a long video such as finding a thief in video of surveillance camera. This study proposes an approach to build an application of recognizing and detecting human in videos. The purpose of this application is to trace, investigate, or review the events that have taken place from the security camera when the need arises. In this work, we have compared there methods (pre-trained Yolov7, self-defined sequential model, and VGG16 Transfer Learning) for detecting human in the video. Experiments are built using data set which is extracted from security camera with several angles and different resolutions. The approach has achieved results with the highest accuracy of 97% for human recognition, thus, it would be a good solution for practice.

Keywords Human detection · Object recognition · Image classification · Security surveillance systems · Yolov7

1 Introduction

An important task for security surveillance systems is human detection in video. When locating a human in a lengthy video, such as locating a burglar (or other people) in a surveillance camera video clips, it might assist shorten the amount of time and effort required. The human detection issue has received a lot of attention in the field of computer vision over the last ten years. In visual content management, tagging (or

N. Thai-Nghe · N. Huu-Hoa (✉)
Can Tho University, 3-2 Street, Can Tho, Vietnam
e-mail: nhhoa@ctu.edu.vn

N. Thai-Nghe
e-mail: ntnghe@ctu.edu.vn

P. Thanh-Son
Vinh Long Radio and Television Station, 50 Pham Thai Buong Street, Vinh Long, Vietnam
e-mail: sonpt@thvl.vn

N. Hoang Phuong et al. (eds.), *Machine Learning and Other Soft Computing Techniques: Biomedical and Related Applications*, Studies in Systems, Decision and Control 543, https://doi.org/10.1007/978-3-031-63929-6_12

identifying) things, particularly people in photos and films, is often a crucial duty. Then, according to requirement, such labeling would enable annotation, search, and retrieval. Finding, recognizing, and keeping an eye on individuals in congested environments and public locations like airports, train stations, and supermarkets is one of the key objectives of video-based surveillance. The ability to identify individuals in security camera footage helps businesses and organizations monitor illicit entry or theft more quickly.

This paper suggests an approach for human detection in video. When necessary, this application's goal is to track down, look into, or review the events captured on security camera. In this work, we have compared the models which are pre-trained Yolov7, self-defined sequential model, and VGG16 Transfer Learning to detect humans. Data sets that are retrieved from security camera footage at various angles and resolutions are used to build experiments.

2 Related Work

From a machine vision standpoint, detecting humans is challenging due of the range of appearances that can be caused by variations in articulation stance, clothes, lighting, and background. However, having an understanding of these constraints can help recognize humans more effectively.

The recognition process usually takes place in two steps: object detection and object classification. Object detection can be performed by background subtraction, optical flow, and space–time filtering. Background subtraction is a common method for object detection where it attempts to detect moving objects from the difference between the current frame and the background frame in a pixel-by-pixel or block-by-block fashion. There are several methods available to perform background subtraction. The most popular models are adaptive Gaussian mixture [1], non-parametric background [2], time difference [3], warped background [4] and hierarchical background model [5]. Optical stream-based object detection [6] uses stream vector characteristics of time-moving objects to detect moving regions in the image sequence. For motion detection based on space–time filtering methods, motion is characterized through the entire volume of three-dimensional (3D) space–time data spanned by the person moving in the image sequences [7].

The three types of object classification techniques are shape-based, motion-based, and texture-based. Shape-based methods begin by outlining the shape details of moving regions like points, boxes, and blobs. Following that, it is typically seen as a typical pattern matching problem. However, because of the human body's flexibility and the various viewing angles, there are many different body forms that could exist, making it challenging to tell one person moving from another. The shape-based method moves items in a different way. Applying department-based pattern matching will help you solve this obstacle [8]. Texture-based methods such as histograms of directed gradients (HOGs) [9] use edge-based dimensional features and use support vector machines (SVMs) to detect regions of children People.

Real-time human detection in aerial video sequences is certainly not without its challenges. For instance, the size of a person detected on aerial video can vary with the height of the unmanned aerial vehicle (UAV). The natural variation in human size detected also poses a challenge to technology [10]. Other issues that arise during acquisition of aerial video include dynamic events such as changes in brightness and noticeable levels of motion blur caused by camera shake [11]. All these issues need to be addressed in order to develop a robust classification method capable of distinguishing humans from non-humans [10]. The CNN has provided machines with the ability to use deep learning (DL) to detect objects of various types. Among the most widely used algorithms for these tasks are Fast R-CNN, Faster R-CNN, Mask R-CNN [14], R-FCN, SqueezeDet [15], EfficientDet [16], MobileNetV2 [17], RetinaNet [18], ShuffleNet [19] and PeleeNet [20].

In this work, we suggest employing the pre-trained deep learning model to detect human in the security camera video for automatically finding a person who appear in the video clip. This approach could help saving time and effort when detecting a human, since it can show and extract the human at a specific time in a long video instead of manually forward and backward by hand.

3 Proposed Approach

The proposed approach is presented in Fig. 1. In this work, the video was extracted from the security camera, then it will be put into the system for analysis. Before detecting, the video will be decomposed into a sequence of frames. Each frame is an image that will then be fed into the classification model (e.g., Yolov7 in this case) to determine human or not_human. However, a long video will have a lot of frames, we will have a parameter of video-stride to determine the number of frames to be skipped (not considered), then, the detection time for the whole video will take place faster.

To compare the classification models for selecting in building the application, in this work we have used three models which are the pre-trained Yolov7, a self-defined sequential model as a baseline, and the VGG16 with Transfer Learning model. Each of these method is described in the following.

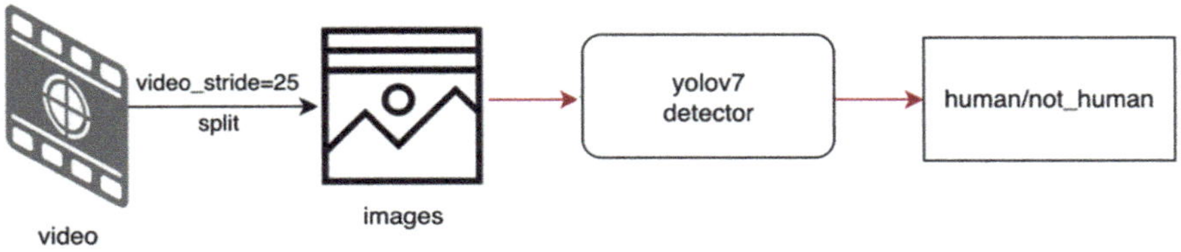

Fig. 1 General architecture of the human detection in video for security surveillance system

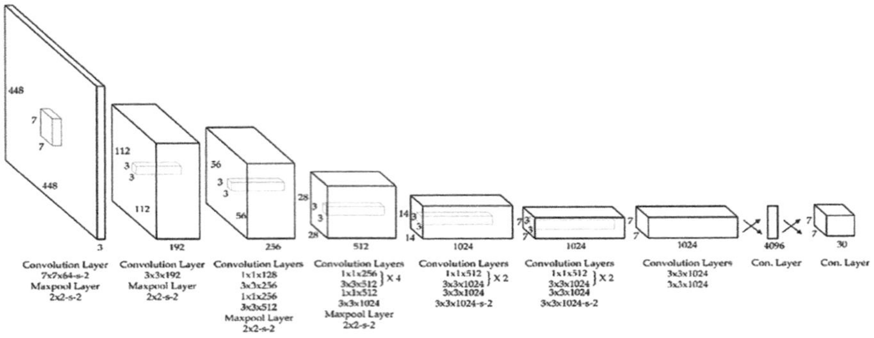

Fig. 2 Yolo architecture [12]

3.1 Yolov7 Detector

The YOLO model [12] consists of two main parts: a deep neural network for feature extraction and a classifier for predicting object classes and bounding box values. The architecture of deep neural networks in YOLO is based on a convolutional neural network (CNN). This neural network has many convolutional layers to extract features from the image. Then, these features will be put into fully connected layers to reduce the number of dimensions and extract important information for object classification. The YOLO's classifier consists of a fully connected class to classify the feature and give probabilities for each class, and a bounding box layer to predict the coordinates and size of the bounding box. Yolo has a total of 24 convolution layers, 4 max-pooling layers and 2 fully connected layers as Fig. 2.

3.2 Sequential Model

A self-defined Sequential model consists of 4 blocks including 4 convolution layers, 4 maximal synthesis layers, and 2 dropout layers Fig. 3. The input layer will receive an image with a resolution of 150 × 150 pixels. The output layer is vector with 2 values corresponding to 2 classes, **human** and **not_human**.

3.3 VGG16 Transfer Learning

The VGG16 [13] network consists of small convolutional filters. The VGG16 has 3 fully connected layers and 13 convolutional layers as in Fig. 4. The full-connected layer with 1,000 labels, while our target dataset has only two classes, human and

```
Model: "sequential"
_________________________________________________________________
 Layer (type)                Output Shape              Param #
=================================================================
 conv2d (Conv2D)             (None, 148, 148, 32)      896

 max_pooling2d (MaxPooling2D  (None, 74, 74, 32)       0
 )

 conv2d_1 (Conv2D)           (None, 72, 72, 64)        18496

 max_pooling2d_1 (MaxPooling  (None, 36, 36, 64)       0
 2D)

 conv2d_2 (Conv2D)           (None, 34, 34, 128)       73856

 max_pooling2d_2 (MaxPooling  (None, 17, 17, 128)      0
 2D)

 conv2d_3 (Conv2D)           (None, 15, 15, 256)       295168

 max_pooling2d_3 (MaxPooling  (None, 7, 7, 256)        0
 2D)

 dropout (Dropout)           (None, 7, 7, 256)         0

 flatten (Flatten)           (None, 12544)             0

 dense (Dense)               (None, 128)               1605760

 dropout_1 (Dropout)         (None, 128)               0

 dense_1 (Dense)             (None, 256)               33024

 dropout_2 (Dropout)         (None, 256)               0

 dense_2 (Dense)             (None, 2)                 514

=================================================================
Total params: 2,027,714
Trainable params: 2,027,714
Non-trainable params: 0
_________________________________________________________________
```

Fig. 3 Self-defined sequential model

not_human. So we'll use this model but "cut" the full-connected layer—aka "top layer". The output is defined as 2 classes corresponding to **human** and **not_human**.

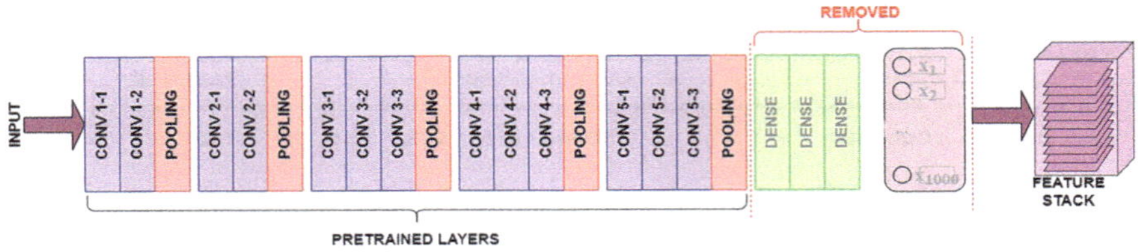

Fig. 4 The VGG16 transfer learning

4 Experiments

4.1 Data Set

The data set[1] is divided into 2 subsets, the training set (with 90% ratio) and and the test set (with 10% ratio) with different resolutions, collected from several sources such as YouTube, indoor and outdoor security cameras. The training set consists of 2 classes: **human** with 500 images and **not_human** with 326 images. The test set consists of 2 classes: **human** with 50 images and **not_human** with 36 images as presented in Fig. 5.

Fig. 5 Sample images in the dataset

[1] https://www.kaggle.com/datasets/constantinwerner/human-detection-dataset.

4.2 Experimental Results

The pre-trained Yolov7 was already trained, we just need to apply it and draw the bounding box when detecting human. Two other models which are the self-defined Sequential model and the VGG16 Transfer Learning need to be trained. The model accuracy and loss of these two models are presented in Figs. 6 and 7.

Confusion matrix of the Yolov7 model is presented in Fig. 8. The VGG-16 Transfer Learning model is presented in Fig. 9 (left side) and the self-defined Sequential model is presented in Fig. 9 (right side). These results show that the Yolov7 has achieved results with the highest accuracy of 97% for human detection in video. Two other models have achieved lower results, 68 and 83% for accuracy for the Sequential and VGG16 Transfer Learning models, respectively. Thus, in this work, we have selected the Yolov7 model for human detection in video.

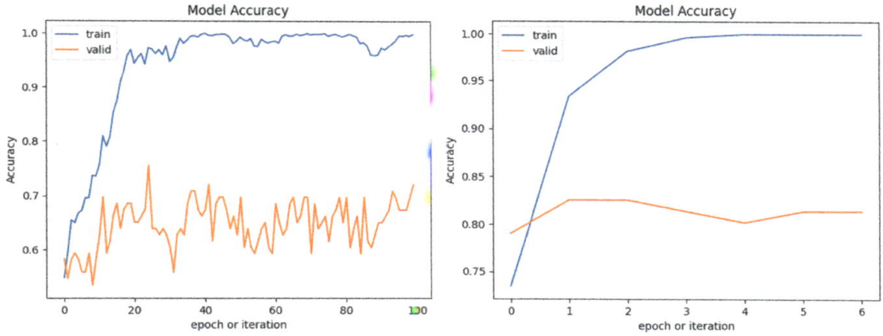

Fig. 6 Training accuracy of the self-defined sequential (left side) and the VGG transfer learning (right side)

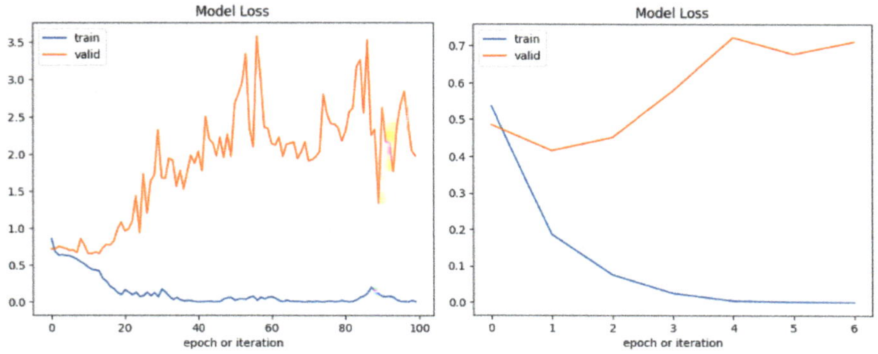

Fig. 7 Training loss of the self-defined sequential (left side) and the VGG transfer learning (right side)

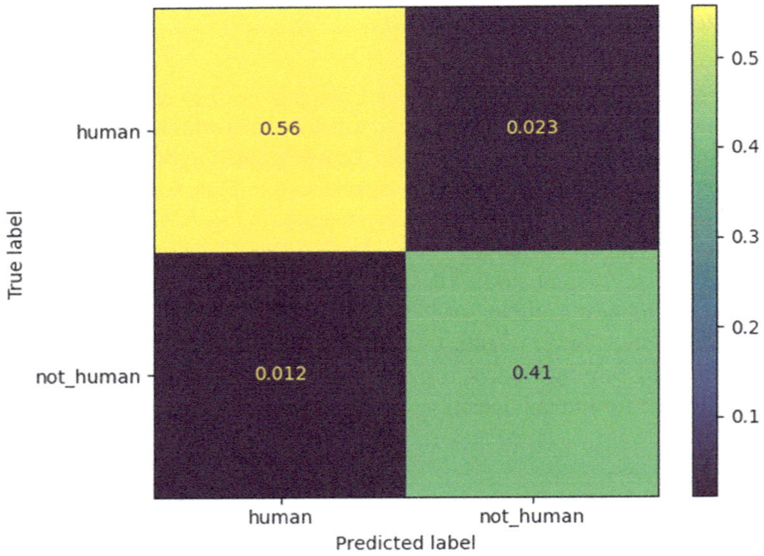

Fig. 8 Confusion matrix of the Yolov7 model

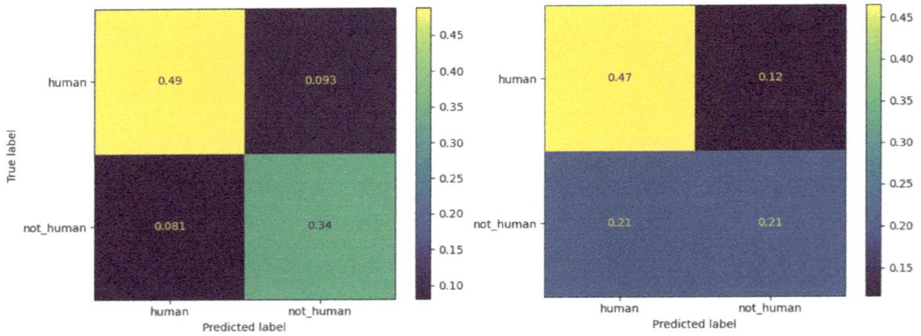

Fig. 9 Confusion matrices of the VGG-16 transfer learning (left side) and the self-defined sequential model (right side)

Figure 10 is a screenshot of the human detection application. After detecting human in a video, the system draws a bounding box and automatically extracts those images and the times where the human appeared in the video.

Figures 11 and 12 present for screenshots of the human detection application when using the self-defined sequential and the VGG-16 Transfer Learning, respectively.

When something was happened, e.g., a thief, instead of tracking by hand in a long video, this system can automatically extract the images and the times where the human appeared in the video. Thus, this could help the users reducing time and effort when finding a person in a long video.

Fig. 10 Human detection (with bounding boxes) in video using Yolov7

Fig. 11 Human detection in video using self-defined sequential model

Fig. 12 Human detection in video using the VGG-16 transfer learning

5 Conclusion

Human detection in video of security surveillance systems could help reducing time and effort when tracking a human in a long video such as finding a thief in the video. In this study, we proposed an approach to build an application of detecting human in videos. The purpose of this application is to trace, investigate, or review the events that have taken place from the security camera when the need arises. There methods (the pre-trained Yolov7, self-defined sequential model, and the VGG16 Transfer Learning) have been compared for detecting human in the video. Experiments were built using data set which is extracted from security camera with several angles and different resolutions. The approach of using Yolo-v7 has achieved results with the highest accuracy of 97% for human recognition, thus, it would be a good solution for practice [21].

Besides, to improve the system in the future, several works could be done, e.g., (i) expand data set to increase model training accuracy and reduce data loss; (ii) adjust hyper-parameters to avoid overfitting; (iii) apply to identify people in real time camera.

References

1. Stauffer, C., Grimson, W.: Adaptive background mixture models for real-time tracking. In: IEEE Conference on Computer Vision and Pattern Recognition (CVPR 1999), pp. 246–252 (1999)
2. Heikkilä, M., Pietäikinen, M.: A texture-based method for modeling the background and detecting moving objects. IEEE Trans. Pattern Anal. Mach. Intell. **28**, 657–662 (2006)
3. Li, L., Huang, W., Gu, I.Y.-H., Tian, Q.: Statistical modeling of complex backgrounds for foreground object detection. IEEE Trans. Image Process **13**, 1459–1472 (2004)
4. Ko, T., Soatto, S., Estrin, D.: Warping background subtraction. In: 2010 IEEE Conference on Computer Vision and Pattern Recognition (CVPR 2010), pp. 1331–1338 (2010)
5. Chen, S., Zhang, J., Li, Y., Zhang, J.: A hierarchical model incorporating segmented regions and pixel descriptors for video background subtraction. IEEE Trans. Ind. Inform 118–127 (2012)
6. Xiaofei, J., Honghai, L.: Advances in view-invariant human motion analysis: a review. IEEE Trans. Syst. Man Cybern. Part C Appl. Rev. 13–24 (2010)
7. Zhong, H., Shi, J., Visontai, M.: Detecting unusual activity in video. In: 2004 IEEE Computer Society Conference on Computer Vision and Pattern Recognition (CVPR 2004), pp. 819–826 (2004)
8. Singh, M., Basu, A., Mandal, M.K.: Human activity recognition based on silhouette directionality. IEEE Trans. Circuits Syst. Video Technol. **18**(9), 1280–1292 (2008)
9. Dalal, N., Triggs, B.: Histograms of oriented gradients for human detection. In: IEEE Conference on Computer Vision and Pattern Recognition (CVPR 2005), pp. 886–893 (2005)
10. AlDahoul, N., Sabri, A.Q.M., Mansoor, A.M.: Real-time human detection for aerial captured video sequences via deep models. Comput. Intell. Neurosci. 1–14 (2018)
11. Bertinetto, L., Valmadre, J., Golodetz, S., Miksik, O., Torr, P.: Staple: complementary learners for real-time tracking. In: Proceedings of the IEEE Conference on Computer Vision and Pattern Recognitio (CVPR), pp. 1401–1409 (2016)
12. Wang, C.-Y., Bochkovskiy, A., Liao, H.-Y.: YOLOv7: Trainable bag-of-freebies sets new state-of-the-art for real-time object detectors (2022). https://doi.org/10.48550/arXiv.2207.02696
13. Liu, S., Deng, W.: Very deep convolutional neural network based image classification using small training sample size. In: 2015 3rd IAPR Asian Conference on Pattern Recognition (ACPR), pp. 730–734. Kuala Lumpur, Malaysia (2015). https://doi.org/10.1109/ACPR.2015.7486599
14. He, K., Gkioxari, G., Dollár, P., Girshick, R.: Mask R-CNN (2022). arXiv:1703.06870
15. Wu, B., Wan, A., Iandola, F., Jin, P.H., Keutzer, K.: SqueezeDet: Unified, small, low power fully convolutional neural networks for real-time object detection for autonomous driving. In: Proceedings of the IEEE Conference on Computer Vision and Pattern Recognitio Workshops (CVPRW), pp. 446–454 (2017)

16. Tan, M., Pang, R., Le, Q.V.: EfficientDet: Scalable and efficient object detection (2022). arXiv: 1911.09070
17. Sandler, M., Howard, A., Zhu, M., Zhmoginov, A., Chen, L.-C.: MobileNetV2: Inverted residuals and linear bottlenecks. In: Proceedings of IEEE/CVF Conference on Computer Vision and Pattern Recognitio, pp. 4510–4520 (2018)
18. Lin, T.-Y., Goyal, P., Girshick, R., He, K., Dollár, P.: Focal loss for dense object detection (2022). arXiv:1708.02002
19. Zhang, X., Zhou, X., Lin, M., Sun, J.: ShuffleNet: an extremely efficient convolutional neural network for mobile devices (2022). arXiv:1707.01083
20. Wang, R.J., Li, X., Ling, C.X.: Pelee: a real-time object detection system on mobile devices (2022). arXiv:1804.06882
21. Thai-Nghe, N., Nguyen, H.H., Park, W., Ngo, Q.T., Truong, M.T.: Human intrusion detection for security cameras using YOLOv8. In: Thai-Nghe, N., Do, TN., Haddawy, P. (eds.) Intelligent Systems and Data Science. ISDS 2023. Communications in Computer and Information Science, vol. 1950. Springer, Singapore (2024). https://doi.org/10.1007/978-981-99-7666-9_19

Fake Face Detection with Separable Convolutions

Hai Thanh Nguyen, Dat Tien Nguyen, Thien Thanh Tran, Huu-Hoa Nguyen, and Nguyen Thai-Nghe

Abstract Fake images bring fake news that causes harmful consequences for victims, many dangerous effects on society, and causes economic damage. For instance, a fake face embedded into an image can be perilous because it can deceive and mislead people, leading to false identification, impersonation, and even fraudulent activities. However, with great support from information technology, such fake images can be easily created by embedding and retouching the victim's face so sophisticated that it is difficult to distinguish with the naked eye. This study proposes approaches to detect fake faces in images using well-known convolutional neural networks, including Separable Convolution-based architectures, Inception, EfficientNet, MobileNet, and Xception, to perform fake face detection in images. The study is carried out on datasets of about 20,000 images with deep learning techniques. Experimental results reveal that separable convolution-based architectures have performed best and better than some previous studies.

1 Introduction

In recent years, with the outstanding development of modern science and technology, especially in machine learning, numerous technologies incorporating Artificial Intelligence (AI) were born to assist people in most specializations [1]. Primarily among them is deep fake (an assortment of the terms "deep learning" and "fake") technology [2], which has received vast controversy from proficients for its potential benefits and menaces. Fundamentally, this study can comprehend that a deep fake is

H. T. Nguyen · D. T. Nguyen · T. T. Tran · H.-H. Nguyen · N. Thai-Nghe (✉)
Can Tho University, Can Tho, Vietnam
e-mail: ntnghe@cit.ctu.edu.vn

H. T. Nguyen
e-mail: nthai.cit@ctu.edu.vn

H.-H. Nguyen
e-mail: nhhoa@ctu.edu.vn

N. Hoang Phuong et al. (eds.), *Machine Learning and Other Soft Computing Techniques: Biomedical and Related Applications*, Studies in Systems, Decision and Control 543, https://doi.org/10.1007/978-3-031-63929-6_13

a photo, audio, or video manipulated by Machine Learning and AI to make it appear to be something it is not.

Deepfakes benefit individuals considerably and help improve work performance in occupations such as film, digital media, education, and healthcare [3]. Specifically in the film industry, applying deep fake to the filming process helps the shooting procedure reach more smoothly, helping to carry out the main actors' scenes even when they are not present on the set. Big stars can also occur in practically any movie they want using deepfake technology. An illustration of using deepfakes in film is the realist aesthetics of digital de-aging, which makes actors appear younger on screen than they are. It has taken an important place in contemporary Hollywood cinema [4]. It has been used in prevalent films such as Gemini Man (produced by director Ang Lee in 2019) and The Irishman (produced by director Martin Scorsese in 2019). To further illustrate the great usefulness of deepfakes, this technology can bring significant benefits in knee osteoarthritis X-rays [5] from generative adversarial neural networks (GANs), provide sweetened capabilities for automatic classification, and make a practical contribution to world medicine. Besides, deepfakes can also obtain a substantial change in human-to-human transmission by helping individuals from multiple different countries with different languages talk directly through the screen of electronic devices such as laptops and smartphones by translating speech, and at the same time, change the movements of the face and mouth to aid communication. Additionally, deepfakes can be used to bring historical figures into the classroom, Thus making the lesson more fascinating. A composite re-enactment will significantly impact pupils. It can raise concentration and be a good wisdom instrument. Hence, this work cannot deny the significant advantages this technology brings to civilization in general and enterprise sectors in particular, especially in the era of vital development of the internet and AI. Nevertheless, deepfakes have an optimistic and a pessimistic side, like other technologies. Escorting the application of deep fake technology to areas of social life is a hazard that can get out of authority. In terms of technology, with breakthroughs in Machine Learning, constructing a bogus video, image, or audio via deepfakes has become highly effortless with several support tools like DeepFace Lab [6], Reface [7], and Faceswap [8]. Along with that, the rapid expansion of GAN deep fake models has created unchecked fake content on social media [9] with malicious purposes targeting individuals and organizations or even influencing the political activities of a country.

Deepfakes have attracted numerous researchers worldwide to discover solutions in the past few years, continuously improving algorithms to detect bogus photos, videos, and audio recordings. For instance, The DeepFake Detection Challenge (DFDC), constructed by Microsoft, Facebook, AWS, AI Partnership, and Scholars, was advanced on Kaggle with a reward of $1,000,000. The contest attracted over 2,000 participants and generated over 35,000 deep fake detection models. Since then, it has shown the essence of deep fake detection for guiding technology companies and researchers' great interest in this topic. Therefore, the advantages and threats mentioned above and the present social concerns. Therefore, developing deep learning models applying AI to detect deepfakes is truthfully urgent and carries tremendous value to the community, especially in protecting the privacy of each individual in the era of fake news [10].

With the development of several convolutional neural networks, image classification performance has obtained significant improvement. However, their efficiency is only partially evaluated in fake face detection. To address this gap, the study conducted a reference process and reviewed different approaches from previous studies, proposing a transfer learning approach that leverages famous pre-trained models, such as Inception, VGG, MobileNet, and Xception, to transfer knowledge to the target domain. Such architectures have achieved very high performance on natural images with 1,000 classes of ImageNet. However, the benefits of these architectures need to be carefully evaluated on fake face detection in images [11]. This approach aims to improve the efficiency of the fake face detection process compared to previous studies.

This paper presents transfer learning techniques to detect fake faces in images effectively. Through the experimental process, the research proves the effectiveness of modern deep learning architectures, with accuracy on the dataset we use reaching up to over 96%. In addition, with testing on numerous different architectures, this article can become a premise to support further research in exploring and expanding the architecture or improving the structure of the current method. This will help improve performance in future fake face detection assignments.

The paper structure consists of 5 parts. First, this study provides an overview of deep fake technology, the advantages and disadvantages of deep fake technology, and the contributions that deep fake technology brings are beneficial in Sect. 2. Second, Sect. 3 introduces the datasets and presents how the proposed method works. Third, the Experimental Sect. 4 demonstrates the offered techniques' experiments. Finally, Sect. 5 will summarize the work and the main contributions.

2 Related Work

Deepfakes have attracted considerable researchers worldwide to discover solutions in the last few years, constantly improving algorithms to detect bogus photos, videos, and audio recordings. With the advent and development of deep learning, this technology shows complete superiority compared to traditional machine learning models, especially in the era of internet development, which entails a considerable amount of uploaded images and videos. Definitive manual identification needs to meet the demand sufficiently. As a result, applying deep learning models makes classification more speedy and leisurely. In this way, performance gradually improves over a period. Following this movement, a wide range of researchers have applied deep learning architectures to detect deepfakes, with some well-known solutions that convey high efficiencies, such as Convolutional Neural Network (CNN) algorithm [12], Recurrent Neural Networks (RNNs) [13], and Long Short-Term Memory (LSTMs)

[14]. These deep learning techniques have been applied in numerous scientific articles and have fetched the predicted outcomes [15–17]. Along with the enthusiastic dedication of researchers, leading technology corporations and organizations also show curiosity, sponsorship, and various ways of encouraging research. To illustrate, Google has discharged an extensive database of deep fake videos, employing paid performers, created to help systems detect affected videotapes that could be used to disseminate misinformation. Besides, deep fake competitions are also held in China, the US, Singapore, and European countries with enormous datasets to attract candidates. Some contests can be mentioned, such as Deeper Forensics Challenge 2020 [18] (*DeeperForensics-1.0* dataset), Deepfake Game Competition 2021 [19] (*Celeb-DF* dataset), and ForgeryNet [20] operates a massive eponymous dataset. Similarly, experiments based on state-of-the-art datasets have been born with considerable research. As proof, a study that opposed Deepfake by Residual Noise Using CNNs [15], this analysis used *FaceForensics++* low-resolution video succession and high-resolution video from DFDC shows that the results yield binary detection accuracy on the *FaceForensics* dataset with 86.8% and *DFDC* with 93%. Moreover, in another work in [21], the authors combined shallow CNNs with long short-term memory to detect fake faces. Hence, the research and curtain-raiser of new approaches to support fake news detection are more sustained by available papers and immense datasets.

Furthermore, study [22] presented an adaptive manipulation trace extraction network (AMTEN). From there came the AMTENnet fake face detector, built by incorporating AMTEN with CNN. When glimpsing fake face images using eclectic FIM techniques, AMTENnet gained an average exactness of up to 98.52%, superior to contemporary technologies. When detecting face photos with untried post-processing operations, the detector earned a moderate accuracy of 95.17%. Besides, research [23] with a different approach has proposed a new detection technique that is a forensic analogy. This approach evaluates the realism of face video frames by detecting the distinction in the similarity between the face area and the background area. To evaluate, the investigation performed training and testing on the *FaceForensics++* dataset and estimated the abstraction ability of the *Celeb-DF* dataset. The experimental outcomes show that the proposed method performs more acceptably or equivalently, specifically in terms of generalization.

Hitherto, multiple investigations and experiments have been conducted, from applying and improving existing methods to developing new ways to detect fake images and videos with greater accuracy and trustworthiness. For example, the study in [24] proposed neural network-based classifiers to witness pseudo-human faces created by both appliances and humans. Their initial results indicate that their method detects GAN-generated images with 94% and human-created images with 74.9% AUROC scores. Meanwhile, the research in [25] presented a new framework to detect fake videos via transfer learning in autoencoders and a hybrid model of CNNs and RNNs. Overall, current image and video tamper detection models offer exceptionally high efficiency; most reach over 70%, and the accuracy gradually improves over time, showing a satisfactory movement in preventing and detecting Deepfakes.

3 Methods

During AI development, besides the contribution of AI in general, deep fake in particular, the deep-fake technique has positives and negatives if any individual uses that technique for bad intentions. However, modern tools such as TensorFlow, PyTorch, and Keras have increasingly strong support for identifying and detecting videos and images created by deep fake technology.

This section generally represents the methods to detect faces in images, as shown in Fig. 1. Firstly, this study must prepare a dataset. This deconstruction uses the *MesoNet Deepfake* dataset to conduct the experimental process. Then, the dataset is used for training with architectures such as InceptionV3 [26], MobileNet [27], VGG, Xception [28] and Separable Convolution-based architecture (our proposal network). We found that the Separate convolution architectures gave very effective results, so we also proposed the architecture for the fake face classification task.

3.1 Dataset

The study collected images from *MesoNet Deepfake* dataset [29] to support fake face detection tasks, as shown in Table 1 with some illustrations including 128×128 images in Fig. 2.

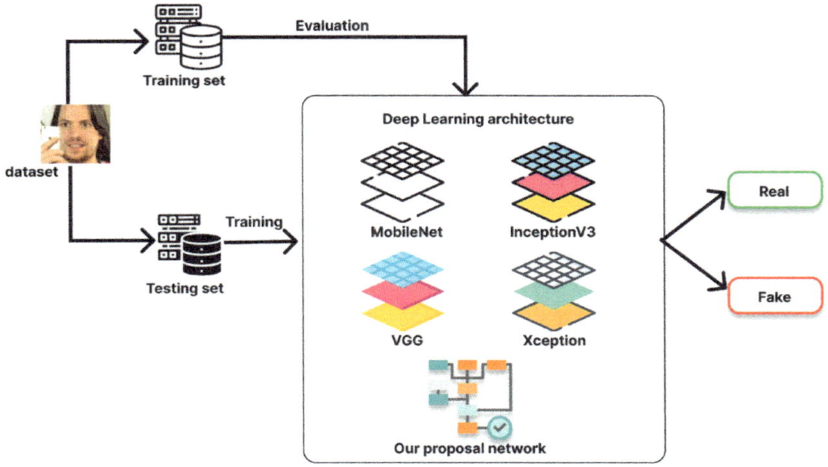

Fig. 1 The proposed workflow for fake image detection

Table 1 The class contribution of MesoNet Deepfake

Dataset	Class	Training	Testing
MesoNet deepfake dataset	Fake	5111	2889
	Real	7250	4259

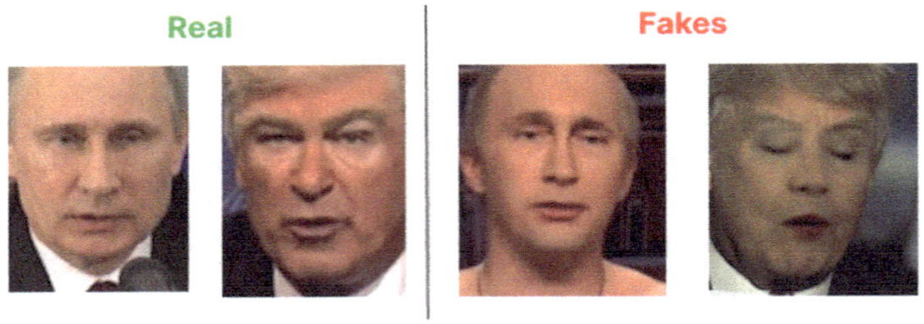

Fig. 2 Some illustrations of real and fake faces in the considered MesoNet dataset

3.2 *Deep Learning Architectures*

This section will detail the architectures used for the Deep Fake detection task, especially the faces generated by modern technology. Figure 3 describes the number of parameters of the proposed models and the size obtained when training the model. It is uncomplicated to notice that the Inception architecture has the most substantial number of params, with 21,819,170 params, followed by the Xception architecture, with 20,872,490 params. Our proposed architecture has the smallest number of parameters at only 1,554,178 params, and the model size is also the smallest at only 17.2 MB. Xception has the most significant model size at 287.8MB, followed by the Inception architecture with 256.8 MB. Compared with Separable Convolutions architectures, our proposed architecture is 13 times smaller than Xception architecture and only half of the MobileNet architecture.

MobileNet [27] was first published in 2017 by Google. The MobileNet is built based on Depthwise Separable Convolutions. MobileNet can reduce the model's parameters through this strategy while maintaining accuracy. The MobileNet structure is conducted from Depthwise Separable Convolutions. The MobileNet

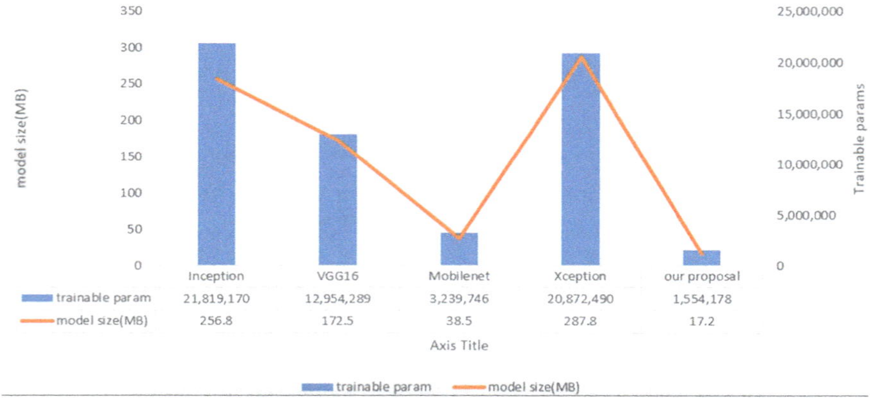

Fig. 3 The number of parameters of the deep learning architectures used in the experiments

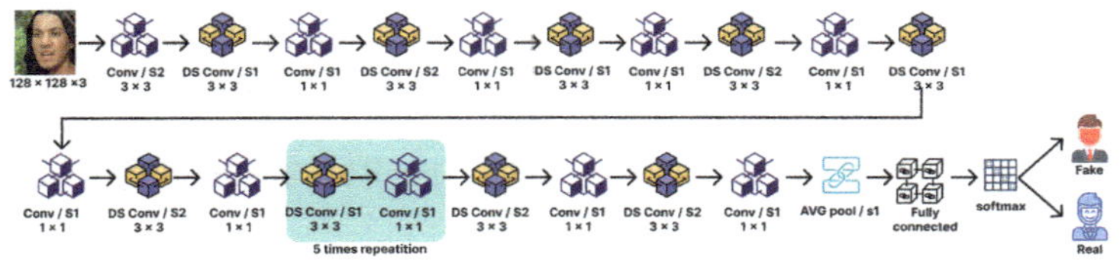

(a) Mobilenet

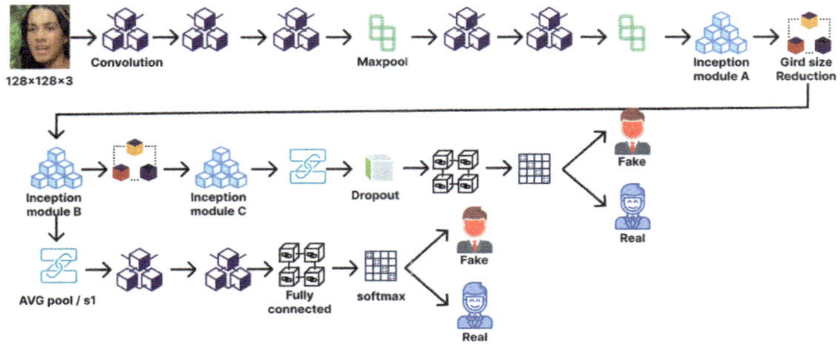

(b) Inceptionv3

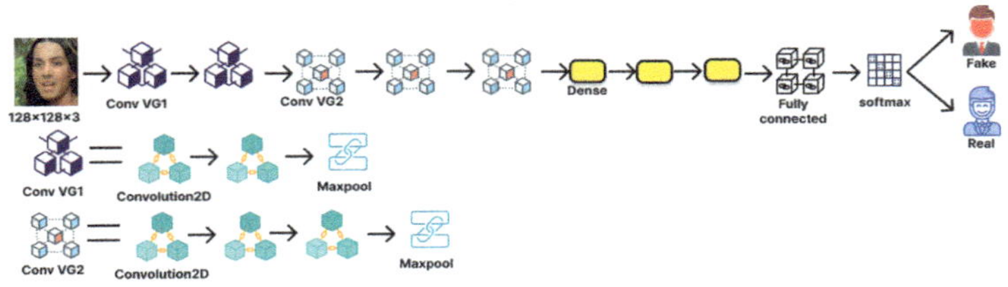

(c) VGG

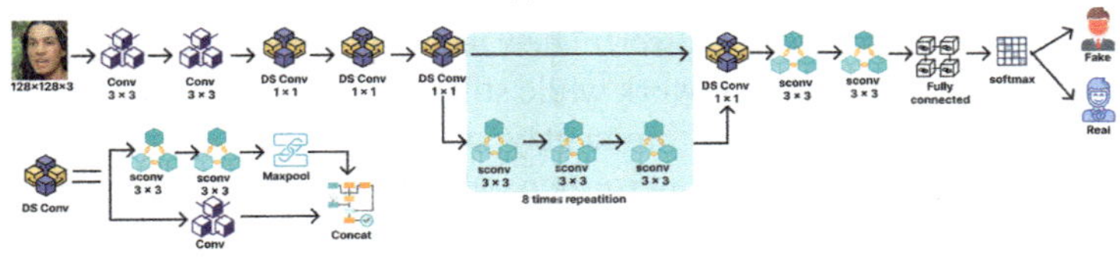

(d) Xception

Fig. 4 Some famous architectures for fake face detection

architecture is constructed on Convolution, Depthwise convolution, and Fully Connected layers. Besides, all layers of MobileNet architecture are followed by a batch norm and ReLU, except the final layer. There are around 6 million trainable parameters in the MobileNet. The detail of MobileNet architecture is portrayed in Fig. 4a.

InceptionV3 [26] architecture was published in 2015. It was built using Convolution, Maxplooing, Dropout, and Concat layers. Layers of the model are observed by batch normalization and a ReLU activation. It was a total of 42 layers, all connected through the Fully Connected layer. InceptionV3 architecture was developed and renovated from InceptionV1 [30] with a total trainable parameter of 25 million. A few renovations of InceptionV3 include Factorization into Smaller Convolutions, Spatial Factorization into Asymmetric Convolutions, Utility of Auxiliary Classifiers, and Efficient Grid Size Reduction. Furthermore, inceptionV3 also resolves representational bottlenecks that InceptionV1 obtains. The details of InceptionV3 architecture are portrayed in Fig. 4b.

VGG [31] is a deep convolutional neural network that was built based on a multilayer idea. VGG16 has a straightforward and deep network architecture. The architecture of VGG16 includes 16 layers, including 13 convolutional layers and 3 fully connected layers. VGG16 has kernel convolution layers (the size of a convolutional region) of a small size of 3×3, and fully connected layers are used for the final classification. The convolutional layers in VGG16 have successive conv2D layers activated by the ReLU function, followed by max-pooling layers with kernel size 2×2 and stride 2 with detailed as shown in Fig. 4c for fake face detection.

Xception [28] was developed by Google as a complex deep Convolution Neural Network architecture that involved Depthwise Separable Convolutions. Inception inspired it, wherein Inception modules have been substituted with depthwise separable convolutions. The first level has a single convolutional level of a 1×1 filter, while the next level has three convolutional levels of a 3×3 filter. This study uses this Xception model to evaluate the distracted driver dataset for classifying the driver's distraction. There are around 24.5 million trainable parameters in the Xception. The details of Xception architecture are depicted in Fig. 4d.

3.3 Separable Convolutions

Separable Convolutions [32] is a powerful new technique that diminishes the number of model parameters in a CNN network while still meeting the requirements of learning image features. There are two prominent types of Separable Convolutions: Spatial Separable Convolutions and Depthwise Separable Convolutions. Spatial Convolution can divide the convolutions into more phases, which helps reduce the number of calculations the model needs to compute. A specific example is as follows: with a 3×3 kernel, it is possible to divide into a 3×1 kernel and a 1×3 kernel. This has the weakness that not all kernels can be split into smaller kernels, which makes it challenging to train the model. The main task of Pointwise Convolution is to increase the number of channels to get the same result as standard Convolution did. 1×1 kernels are used to perform Convolution with each data point. The kernel will have as numerous channels as the number of channels of the input image to get one channel of the output image.

Through testing transfer learning models, we have proposed improved models from Separable Convolutions and obtained specific efficiency. By taking advantage of the godsends of Depthwise Separable Convolutions, we have designed an architecture built from MobileNet and Xception, namely **DSCMX**. Specifically, as shown in Fig. 5, this model is built mainly with the following layers: Batnormaization, Depthwise Separable Convolutions, convolution2D, and globalAvgpool. These

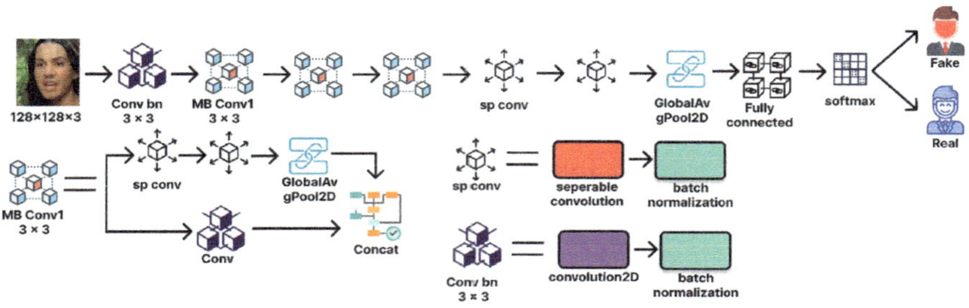

Fig. 5 The proposed architecture builds a base on separable convolutions layer

layers are connected based on each other according to the previously proposed models, and the purpose is to decrease the number of calculation formulas and enhance the model's accuracy. In this way, the model has achieved a specific efficiency when distinguishing deep fake faces.

4 Experimental Results

In this section, we will test the proposed models. Moreover, the environment setting configurations are also shown in detail. The results shown will show the main contribution to our proposed model. In addition, it also shows the effectiveness of the model.

4.1 Environment Setting

In this section, the experimental process of the proposed method will be presented. The testing process used a dataset of around 19,509 images belonging to 2 classes, fake and real, and all of the datasets have been resized down to 128×128. All models are configured with a batch size of 8 and an Adam optimizer with a default learning rate of 0.001, running to 100 epochs. To evaluate the model, all datasets are divided five times according to the hold-out type and training. As mentioned in Table 1, the study split 12,361 samples for training and 7,148 samples for testing on the *MesoNet Deepfake* dataset. All runtime is performed on Colab, equipped with NVIDIA Geforce MX150 2 GB and connected to a personal Core-i7 with 8 GB of RAM. The recommended configuration is shown in Table 2.

4.2 Results

Table 2 summarizes InceptionV3, MobileNet, VGG, Xception, and Our proposal architecture parameters. The architecture proposed is trained for 100 epochs with

Table 2 Configuration for the considered architectures

Hyper-parameters	Value
Input size	128 × 128 × 3
Number of epochs	100
Batch size	8
Activation	ReLU
Optimizer	Adam
Learning rate	0.001

a batch size of 8. So that the training process can converge faster, the study uses RELU activation to evaluate the performance of a model. Besides, the study uses two metrics, including the accuracy score, the area under the curve (AUC) of the receiver operating characteristics (ROC), and the F1-score that metrics measure the classification results of two classes, fake and real. After an experiment on *MesoNet Deepfake* dataset, the study obtains the results shown in Table 3. The consequence from that table performance of the accuracy on the training set from architectures such as Xception, MobileNet, VGG, and InceptionV3 correspond to 99.4, 99.5, 79.3, and 99.3%. The architecture result shows that our proposal network, DSCMX, scored superior to the rest of the architecture. Therein, VGG16 architecture achieved the lowest score. The accuracy score of VGG16 architecture just achieved 79.3 and 80% on AUC scores. Besides, the experiment results on the testing set also show that Separable Convolutions architecture is suitable for fake and real image classification. The metrics such as Loss, Accuracy (ACC), AUC, and F1-score obtained show that Xception is the best architecture corresponding to 12.9, 96.2, 99 and 96.2%. The worst result still belongs to the VGG16 architecture, with an accuracy of only 75.7% and an AUC of only 80.2% (Fig. 7).

Table 3 Training results on mesonet deepfake dataset using architectures such as Xception, MobileNet, InceptionV3, VGG, and the proposed architecture

	Xception	Mobilenet	Inception	VGG	DSCMX
The performance on the training set					
Loss	0.0128(±0.015)	0.0206(±0.017)	0.1932(±0.323)	0.3765(±0.285)	0.0018(±0.001)
ACC	0.9958(±0.004)	0.9948(±0.001)	0.9930(±0.006)	0.7927(±0.190)	0.9995(±0.001)
AUC	0.9996(±0.001)	0.9996(±0.004)	0.9965(±0.005)	0.8215(±0.215)	0.9999(±0.000)
F1-score	0.9958(±0.001)	0.9948(±0.040)	0.9930(±0.006)	0.7922(±0.190)	0.9995(±0.001)
The performance on the testing set					
Loss	0.1745(±0.040)	0.1965(±0.008)	0.4249(±0.431)	0.4835(±0.175)	0.1294(±0.030)
ACC	0.9540(±0.002)	0.9439(±0.001)	0.9123(±0.056)	0.7573(±0.145)	0.9623(±0.002)
AUC	0.9887(±0.002)	0.9827(±0.004)	0.9567(±0.044)	0.8020(±0.185)	0.9897(±0.004)
F1-score	0.9540(±0.002)	0.9439(±0.008)	0.9123(±0.056)	0.7573(±0.145)	0.9623(±0.001)
Time (s)	37.213(±6.35)	18.183(±6.50)	96.345(±22.79)	17.717(±5.25)	47.164(±18.47)

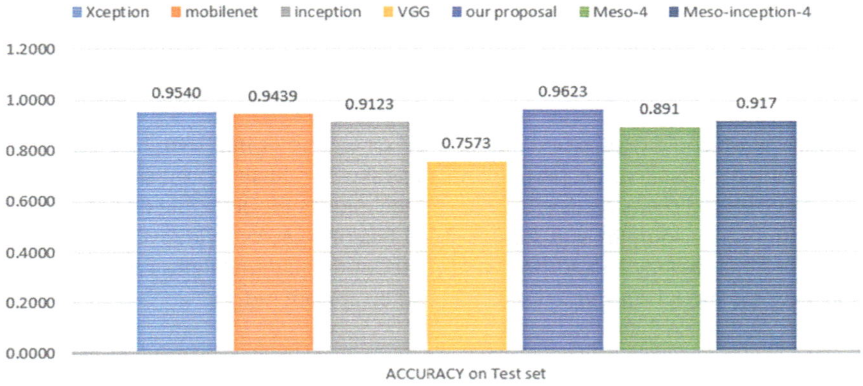

Fig. 6 Result comparison of various architectures on MesoNet Deepfake dataset

Fig. 7 An illustration of some results of the testing set

As compared in Fig. 6 with the Meso-inception-4 [30] and Meso-4 [30], their results are entirely inferior to our proposed method, only about 91.7%, and our method has a higher level of 96.2% on our network model. Because the learning process probably takes longer, the accuracy is also significantly improved, and the learning time is longer, partly because this network model has fewer parameters than other architectures. The shown result reveals that our proposal leverages the advantage of Separable Convolutions and gives superior scores from the rest of the network.

5 Conclusion

The study presents transfer learning techniques to detect fake faces in images based on *MesoNet Deepfake* dataset. It also demonstrates the efficacy of modern, efficient deep learning architectures for deepfake classification tasks. The experiment's results indicated that modern architectures could be beneficial for detecting fake faces in the datasets above. Specifically, Our proposal architecture yielded the highest accuracy score of 96.2% and an AUC score of 99% on the *MesoNet Deepfake* dataset. In contrast, the VGG16 architecture achieved a comparatively lower accuracy score of 75.7%. The accuracy scores achieved on the datasets are significant, and the proposed method is expected to have practical applications across various fields, such as politics, the military, economics, and education.

Future work could explore extending the architecture or improving the structures of Xception to enhance the performance in fake face detection tasks. Moreover, further studies can focus on developing new techniques for detecting deep fakes that can identify more complex and sophisticated fake images and videos by optimizing their hyper-parameters. In addition, the impact of data augmentation techniques on improving the model's accuracy score should be considered.

References

1. Holzinger, A., Keiblinger, K., Holub, P., Zatloukal, K., Müller, H.: AI for life: trends in artificial intelligence for biotechnology. New Biotechnol. **74**, 16–24 (2023). https://www.sciencedirect.com/science/article/pii/S1871678423000031
2. Gil, R., Virgili-Gomà, J., López-Gil, J.M., García, R.: Deepfakes: evolution and trends. Soft Comput. (2023). https://doi.org/10.1007/s00500-023-08605-y
3. Pandey, C.K., Mishra, V.K., Tiwari, N.K.: Deepfakes: when to use it. In: 2021 10th International Conference on System Modeling & Advancement in Research Trends (SMART), pp. 80–84 (2021). ISSN: 2767-7362
4. Loock, K.: On the realist aesthetics of digital de-aging in contemporary Hollywood cinema. Orb. Litt. **76**(4), 214–225 (2021). https://doi.org/10.1111/oli.12302, publisher: Wiley
5. Prezja, F., Paloneva, J., Polonen, I., Niinimaki, E., Ayramo, S.: DeepFake knee osteoarthritis X-rays from generative adversarial neural networks deceive medical experts and offer augmentation potential to automatic classification. Sci. Rep. **12**(1) (2022). https://doi.org/10.1038/s41598-022-23081-4, publisher: Springer Science and Business Media LLC
6. Perov, I., Gao, D., Chervoniy, N., Liu, K., Marangonda, S., Umé, C., Dpfks, M., Facenheim, C.S., RP, L., Jiang, J., Zhang, S., Wu, P., Zhou, B., Zhang, W.: DeepFaceLab: integrated, flexible and extensible face-swapping framework (2020). arXiv:2005.05535
7. Parks, C.L., Richard, A.H., Monson, K.L.: Preliminary performance assessment of computer automated facial approximations using computed tomography scans of living individuals. Forensic Sci. Int. **233**(1–3), 133–139 (2013). https://doi.org/10.1016/j.forsciint.2013.08.031, publisher: Elsevier BV
8. Nawaz, M., Masood, M., Javed, A., Nazir, T.: FaceSwap based DeepFakes detection. Int. Arab. J. Inf. Technol. **19**(6) (2022). https://doi.org/10.34028/iajit/19/6/6, publisher: Zarqa University
9. Tolosana, R., Vera-Rodriguez, R., Fierrez, J., Morales, A., Ortega-Garcia, J.: Deepfakes and beyond: a survey of face manipulation and fake detection. Inf. Fusion **64**, 131–148 (2020). https://doi.org/10.1016/j.inffus.2020.06.014, publisher: Elsevier BV
10. Pennycook, G., Rand, D.G.: The psychology of fake news. Trends Cogn. Sci. **25**(5), 388–402 (2021). https://doi.org/10.1016/j.tics.2021.02.007, publisher: Elsevier BV
11. Sharma, J., Sharma, S., Kumar, V., Hussein, H.S., Alshazly, H.: Deepfakes classification of faces using convolutional neural networks. Trait. du Signal **39**(3), 1027–1037 (2022), publisher: International Information and Engineering Technology Association
12. Teuwen, J., Moriakov, N.: Convolutional neural networks. In: Handbook of Medical Image Computing and Computer Assisted Intervention, pp. 481–501. Elsevier (2020). https://doi.org/10.1016/b978-0-12-816176-0.00025-9
13. Salehinejad, H., Sankar, S., Barfett, J., Colak, E., Valaee, S.: Recent Advances in Recurrent Neural Networks (2018). arXiv:1801.01078

14. Hochreiter, S., Schmidhuber, J.: Long short-term memory. Neural Comput. **9**(8), 1735–1780 (1997). https://doi.org/10.1162/neco.1997.9.8.1735, publisher: MIT Press - Journals
15. Rai, M.C.E., Ahmad, H.A., Gouda, O., Jamal, D., Talib, M.A., Nasir, Q.: Fighting deepfake by residual noise using convolutional neural networks. In: 2020 3rd International Conference on Signal Processing and Information Security (ICSPIS). IEEE (2020). https://doi.org/10.1109/icspis51252.2020.9340138
16. Guera, D., Delp, E.J.: Deepfake video detection using recurrent neural networks. In: 2018 15th IEEE International Conference on Advanced Video and Signal Based Surveillance (AVSS). IEEE (2018). https://doi.org/10.1109/avss.2018.8639163
17. Stanciu, D.C., Ionescu, B.: Deepfake video detection with facial features and long-short term memory deep networks. In: 2021 International Symposium on Signals, Circuits and Systems (ISSCS). IEEE (2021). https://doi.org/10.1109/isscs52333.2021.9497385
18. Jiang, L., Guo, Z., Wu, W., Liu, Z., Liu, Z., Loy, C.C., Yang, S., Xiong, Y., Xia, W., Chen, B., Zhuang, P., Li, S., Chen, S., Yao, T., Ding, S., Li, J., Huang, F., Cao, L., Ji, R., Lu, C., Tan, G.: DeeperForensics challenge 2020 on real-world face forgery detection: methods and results (2021). arXiv:2102.09471
19. Peng, B., Fan, H., Wang, W., Dong, J., Li, Y., Lyu, S., Li, Q., Sun, Z., Chen, H., Chen, B., Hu, Y., Luo, S., Huang, J., Yao, Y., Liu, B., Ling, H., Zhang, G., Xu, Z., Miao, C., Lu, C., He, S., Wu, X., Zhuang, W.: DFGC 2021: a deepfake game competition. In: 2021 IEEE International Joint Conference on Biometrics (IJCB). IEEE (2021). https://doi.org/10.1109/ijcb52358.2021.9484387
20. He, Y., Sheng, L., Shao, J., Liu, Z., Zou, Z., Guo, Z., Jiang, S., Sun, C., Zhang, G., Wang, K., Yue, H., Hong, Z., Wang, W., Li, Z., Wang, Q., Wang, Z., Xu, R., Zhang, M., Wang, Z., Huang, Z., Zhang, T., Zhao, N.: ForgeryNet—face forgery analysis challenge 2021: methods and results (2021). arXiv:2112.08325
21. Nguyen, H.T., Dao, T.C., Phan, T.M.N., Phan, T.T.: Fake face detection in video using shallow deep learning architectures. Int. J. Intell. Syst. Technol. Appl. **20**(6), 469 (2022). https://doi.org/10.1504/ijista.2022.128528
22. Guo, Z., Yang, G., Chen, J., Sun, X.: Fake face detection via adaptive manipulation traces extraction network. Comput. Vis. Image Underst. **204**, 103170 (2021). https://www.sciencedirect.com/science/article/pii/S107731422100014X
23. Pan, Z., Ren, Y., Zhang, X.: Low-complexity fake face detection based on forensic similarity. Multimed. Syst. **27**(3), 353–361 (2021). https://doi.org/10.1007/s00530-021-00756-y
24. Tariq, S., Lee, S., Kim, H., Shin, Y., Woo, S.S.: Detecting both machine and human created fake face images in the wild. In: Proceedings of the 2nd International Workshop on Multimedia Privacy and Security, pp. 81–87. MPS '18, Association for Computing Machinery, New York, NY, USA (2018). https://doi.org/10.1145/3267357.3267367
25. Suratkar, S., Kazi, F.: Deep fake video detection using transfer learning approach. Arab. J. Sci. Eng. (2022). https://doi.org/10.1007/s13369-022-07321-3
26. Szegedy, C., Vanhoucke, V., Ioffe, S., Shlens, J., Wojna, Z.: Rethinking the inception architecture for computer vision (2015). _eprint: 1512.00567
27. Howard, A.G., Zhu, M., Chen, B., Kalenichenko, D., Wang, W., Weyand, T., Andreetto, M., Adam, H.: MobileNets: efficient convolutional neural networks for mobile vision applications (2017). _eprint: 1704.04861
28. Chollet, F.: Xception: Deep learning with depthwise separable convolutions. In: 2017 IEEE Conference on Computer Vision and Pattern Recognition (CVPR). IEEE (2017). Event-place: Honolulu, HI
29. Afchar, D.: MesoNet (2023). https://github.com/DariusAf/MesoNet, original-date: 2018-04-27T07:30:38Z
30. Afchar, D., Nozick, V., Yamagishi, J., Echizen, I.: MesoNet: a compact facial video forgery detection network (2018). _eprint: 1809.00888
31. Simonyan, K., Zisserman, A.: Very deep convolutional networks for large-scale image recognition. In: International Conference on Learning Representations (2015)
32. Chollet, F.: Xception: deep learning with depthwise separable convolutions. In: 2017 IEEE Conference on Computer Vision and Pattern Recognition (CVPR), pp. 1800–1807 (2017)

A Classification System of Mammograms Based on Convolutional Neural Networks

Nguyen Hoang Phuong, Nguyen Duc Dan, Nguyen Viet Dung, Ha Manh Toan, Nguyen Khac Dung, and Dao Van Tu

Abstract This paper presents an approach to improve a classification system based on convolutional neural networks for classifying breast cancer X-ray images into three classes of Normal, Benign, and Malignant by increasing the training set. For training convolutional neural network models, we use 15,040 Vietnamese mammograms, which were collected and annotated by the radiologists of the Vietnam National Cancer Hospital. Our experiment was conducted with the Resnet 34. The evaluation of system performance using the testing set achieves a macAUC of 0.84807, an average sensitivity of 0.69058, and an average specificity of 0.84336 which are higher than a macAUC of 0.828247, an average sensitivity of 0.64738, and an average specificity of 0.825670 when we trained the model Resnet 18 with 8,777 mammograms. For the classification of breast cancer X-ray images into malignant class, our network achieves an AUC of 0.89758. On the other hand, our proposed model achieves results which is higher than the results of a classification system applied to the model ResNet 50 for classifying breast cancer X-ray images into BI-RADS 045, BI-RADS 1, BI-RADS 23 categories which achieves a macAUC of 0.754, an average sensitivity of 0.58, and an average specificity of 0.79 when using

N. Hoang Phuong · N. D. Dan
Artificial Intelligence Division, Information Technology Faculty, Thang Long University, Nghiem Xuan Yem Road, Hoang Mai District, Hanoi, Vietnam

N. V. Dung
Department of Electronics, School of Electrical and Electronic Engineering, Hanoi University of Science and Technology, Hanoi, Vietnam
e-mail: dung.nguyenviet1@hust.edu.vn

H. M. Toan
Institute of Information Technology, Vietnam Academy of Science and Technology, Hanoi, Vietnam
e-mail: hmtoan@ioit.ac.vn

N. K. Dung · D. V. Tu (✉)
Vietnam National Cancer Hospital, 30 Cau Buou, Tan Trieu, Thanh Tri, Hanoi, Vietnam
e-mail: daovantu@hmu.edu.vn

N. K. Dung
e-mail: khacdung.nguyen@bvk.org.vn

N. Hoang Phuong et al. (eds.), *Machine Learning and Other Soft Computing Techniques: Biomedical and Related Applications*, Studies in Systems, Decision and Control 543, https://doi.org/10.1007/978-3-031-63929-6_14

7,912 X-Ray images collected by Vietnam Hanoi Medical University Hospital. We observe that the classification performance improves as the number of mammograms increases. With the achieved result, our classification system can help radiologists in Vietnam in the interpretation of breast cancer X-ray images.

Keywords Convolutional neural network · Classification system · Mammograms · ResNet

1 Introduction

In Vietnam, breast cancer is one of the most frequent cancers of women. Women of all ages after puberty can have this cancer, and the rate would grow over time. There were 21,555 new cases of breast cancer in 2020 in Vietnam, achieving 25.8% with a total of 83,647 new cancer cases for all women of all ages [10]. To reduce the late-stage diagnosis rate of breast cancer, mammography in breast cancer screening is an efficient tool, and it has become more valuable. On the other hand, Vietnam lacks experienced radiologists for breast cancer X-ray interpretation in national and provincial hospitals. In recent years, some intelligent systems for classifying mammograms have been developed. In the work of New York University [4], the authors applied deep neural networks for breast cancer X-ray exam classification to improve radiologists' performance in breast cancer screening. In their work, the authors used over 1,000,000 images for training and testing the system. Their network achieves an AUC of 0.895 for the classification of cancer in the breast. In Vietnam, in work [5], the authors develop a convolutional neural network-based system for classifying breast cancer X-ray images into three classes of BI-RADS categories as BI-RADS 1 ("normal"), BI-RADS 23 ("benign") and BI-RADS 045 ("incomplete and malignance"). This classification system is developed based on the convolutional neural network with ResNet 50. The system is trained and tested on a breast cancer image dataset of Vietnamese patients containing 7,912 images provided by radiologists at the Hanoi Medical University Hospital. The system accuracy uses the testing set achieved a macAUC (a macro average of the three AUCs) of 0.754. In our study [12], the authors used 8,777 breast cancer X-ray images of the patients of the Vietnam National Cancer Hospital for training the model Resnet 18 and the model achieved a macAUC of 0.828247 for the classification of X-ray images into Normal, Benign, and Malignant classes. This measured value for evaluation of the system performance is enough good, but we try to increase this value as much as possible. Therefore, in our study, we improve the performance of the system using the convolutional neural network Resnet 34 to classify X-ray breast cancer images into three classes Normal, Benign, and Malignant [12] by increasing the number of mammograms used for training the model. The main contributions of this paper are:

(1) Radiologists of the Vietnam National Cancer Hospital in Hanoi collect and label 15,040 X-ray images into three classes: Normal, Benign, and Malignant.

(2) We used these images for training the model Resnet 34. The results of the experiment will also reflect the effect of the training sample size.
(3) Comparison of our result with the existing works shows some advantages of the proposed method which achieves the highest AUCs for the classification of Normal, Benign, and Malignant images.

The rest of the paper is organized as follows: Sect. 2 presents a system design. Section 3 describes data collection and labeling. Section 4 describes data preprocessing. Section 5 presents a model training and evaluation. Section 6 gives some conclusions.

2 System Design

A convolutional neural network core model for breast cancer classification based on mammograms is described in Fig. 1. In the previous experiments, we applied different backbone convolutional neural network models (CNNs) such as Resnet 18, Resnet 34, Resnet 50, and VGG 16 and trained these models on 2,101 breast cancer images [12] received from the Vietnam National Cancer Hospital. These models are built based on the architecture described in Fig. 1. We found that the backbone model Resnet 34 is optimal among the above models and chose this model for our experiment in this study.

In Fig. 1, a backbone model is a CNN model. The pooling layer is usually designed after the convolution layer in convolutional neural networks CNNs. It reduces the computation space and also helps to reduce the problem of overfitting. The flatten layer converts 2-dimensional data of the convolution layer to 1-dimensional data of a fully connected layer. Then, data will be calculated with a fully connected layer. Dropout is a technique to disconnect connections randomly during learning. It is a technique to prevent neural networks from overfitting problems [6]. FC 256

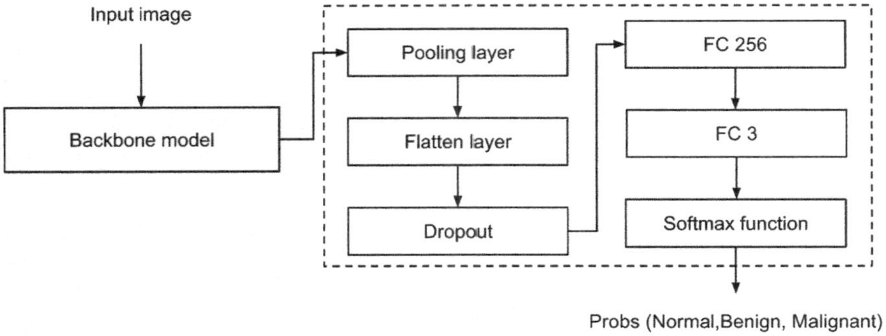

Fig. 1 The CNN architecture

and FC 3 are fully connected layers with sizes 256 and 3. FC 256 is an intermediate computational layer, and FC 3 will generate the final result before entering it into the softmax function. The softmax function keeps the output probabilities in [0, 1], and the sum of the probabilities equals 1. This property is consistent with the conventional understanding of the concept of probability. Three probability values are values representing the input image belonging to the classes of normal, benign, and malignant.

In more detail, the resolution of the input image is 512×512 pixels. The input image will be put into the backbone model. In the experiment, the backbone model is the model Resnet 34. After the backbone model is the additional parts for the inference task. In the implementation, all elements of our CNN architectures are built with the support of the PyTorch library.

3 Data Collection and Labeling

In our study, the radiologists of the Vietnam National Cancer Hospital collected from each patient four breast cancer images using screening mammography as R-CC (right craniocaudal), R-MLO (right mediolateral oblique), L-CC (left craniocaudal), L-MLO (left mediolateral oblique). An example of a breast cancer image of a patient is shown in Fig. 2.

Every breast cancer X-ray image is classified into one of the classes Normal, Benign, or Malignant. Radiologists use the PMR tool of the Vietnam National Cancer Hospital to localize the abnormal border of Benign and Malignant images.

The total of the breast cancer X-ray images collected by radiologists of the Vietnam National Cancer Hospital is 15,040 images. This dataset is divided into three parts:

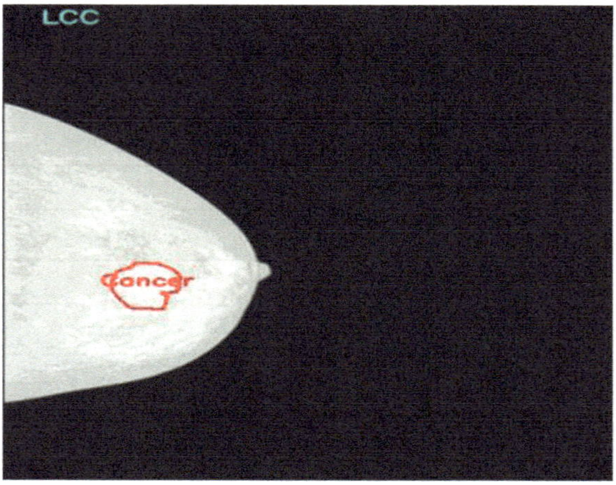

Fig. 2 Example of a screening image L-CC of a patient

the training dataset, the validation dataset, and the test dataset. The dataset is divided by the rate of 80:10:10.

4 Data Pre-processing

Input data is a Dicom X-ray image of breast cancer. The process aims to improve the quality of input images for training the deep learning network classification model. The received mammogram has different standard views used in screening mammography, for example, R-CC and L-CC or R-MLO and L-MLO. Besides, the breast only occupies a part of the image area. Therefore we must flip the image for the same breast orientation and perform a crop to get an image with an area of most of the breast.

The main steps of image pre-processing are as follows:

Step 1: We first flip along the vertical axis the images of the right breast R-CC and R-MLO before feeding them to the training process so that all the breast images are rotated in the right direction.

Step 2: We check the mammograms received from the radiologists. If there are letters on the mammogram, for example, L-CC, R-CC, L-MLO, R-MLO, or other noise, we will remove them.

Step 3: In the received mammograms, the breast part occupies only a part of the image, so we crop as much as possible the black area (in Fig. 2) to get a breast image.

Step 4: This image is resized to 512×512 pixels.

Step 5: Finally, each image is normalized.

After the pre-processing steps, all images are used for the model training.

5 Model Training and Evaluation

The server installed OS Linux Ubuntu 16.04.3 LTS with libraries that support deep neural network algorithms on GPUs containing Nvidia CUDA. The breast cancer X-ray classification system has been developed on Pytorch.

The learning rate reducing on the plateau was used during training with 5e–3 of the initial learning rate and 1e–8 of the minimum learning rate. It is a technique to reduce the learning rate when the error value does not decrease during the learning process. The model training consists of two phases.

In model training, we use the Resnet 34 model applying the weighted loss function, but we don't establish an approximately equal number of samples of classes of Normal, Benign, and Malignant, and we don't apply gray-level pre-processing techniques [3, 7, 8]. In the experiment, three metrics as a macAUC, an averaged sensitivity, and an averaged specificity are used for the evaluation of Resnet 34 trained on the dataset described in Table 1.

Table 1 Dataset for training

	Malignant	Normal	Benign	Total
Training set	2534	6349	3145	12,028
Validation set	316	794	393	1503
Test set	319	796	394	1509
Total	3169	7939	3932	15,040

The graph in Fig. 3 depicts the variation of the loss function values calculated in the training set and the validation set during training. The graph shows that the training process reaches the optimal point at epoch 33. After this point, the validation loss value shows no obvious signs of decline. Furthermore, the validation accuracy value peaks at epoch 33. Therefore, the model at epoch 13 is optimal and this model is used for the experiment on the test set.

Table 2 shows that the macAUC obtains the value of 0.84807. This is the macro average of three AUCs: AUC of Malignant versus others is 0.89758, AUC of Normal versus others is 0.86055, AUC of Benign versus others is 0. 78,608. Similarly, the model achieves an average Sensitivity value of 0.69058 and an average Specificity value of 0.84336. In Table 2, from our point of view, the AUC of Malignant versus others achieves the highest accuracy with the value of 0.89758 because the feature structure of cancer images is more "visible" than the feature structures of Benign and Normal X-ray images. In the clinical practice of breast cancer diagnosis, radiologists expect to have a high accuracy in the diagnosis of cancer.

Table 3 presents a confusion matrix of the test dataset with the model based on the Resnet 34 backbone.

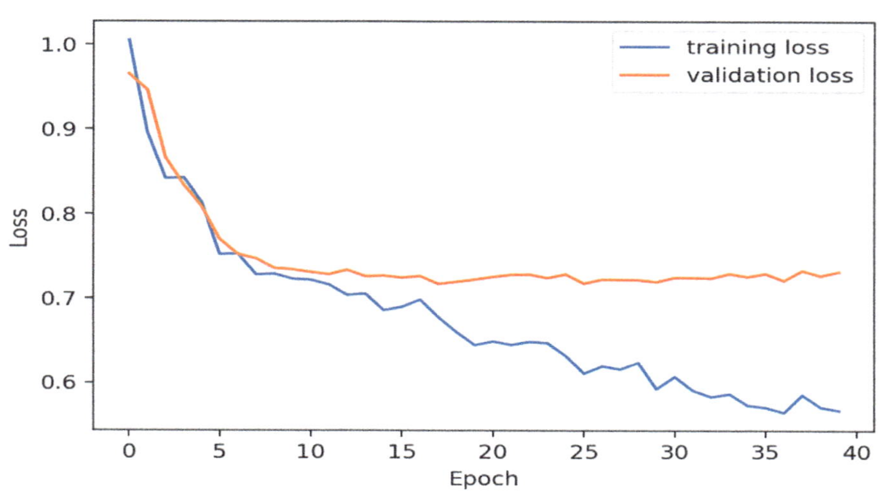

Fig. 3 Graph of loss values of train set and validation set by epochs

Table 2 The model's results

Prediction hypotheses	(mac) AUC	Sensitivity	Specificity
Malignant versus others	0.89758	0.75235	0.92521
Normal versus others	0.86055	0.78894	0.75736
Benign versus others	0.78608	0.53045	0.84753
macAUC/Average	0.84807	0.69058	0.84336

Table 3 The confusion matrix

True label	Malignant	240	42	37
	Normal	35	628	133
	Benign	54	131	209
		Malignant	Normal	Benign
		Predicted label		

The result of Table 3 is described as follows:

- Among the 319 images labeled Malignant, the system correctly recognized 240 images corresponding to a rate of 75.24%, 42 images were mistakenly identified as images labeled Normal with a rate of 13.17%, and 37 images were mistakenly identified as images labeled Benign corresponding to the rate of 11.60%.
- Among the 796 images labeled Normal, the system correctly recognized 628 images corresponding to a rate of 78.89%, 35 images were mistakenly identified as images labeled Malignant corresponding to 4.40%, and 133 images were mistakenly identified as images labeled Benign corresponding to the rate of 16.71%.
- Among the 394 images labeled Benign, the system correctly recognized 209 images corresponding to a rate of 53.05%, 54 images were mistakenly recognized as Malignant images corresponding to a rate of 13.71%, and 131 images were mistakenly recognized as Normal labels corresponding to the rate of 33.25%.

Table 4 shows the comparison of the result of our model with the related studies in [5, 12].

In Table 4, we observe that the proposed Resnet 34 model used 15,040 breast cancer X-Ray images collected by the Vietnam National Cancer Hospital for model training and achieved a macAUC of 0.84807 for the classification of Normal, Benign, and Malignant images which is higher than the accuracy of the Resnet 18 model used only 8,777 breast cancer X-Ray images for training model. This Resnet 18 model achieves only a macAUC of 0.828247 for the classification of Normal, Benign, and Malignant images. This comparison shows that even though we increased the dataset for training the Resnet 34 model almost two times with 15,040 images in comparison with 8,777 images of the Resnet 18 model, the macAUC accuracy of the Resnet 34 model increases only a few as 0.01983.

Table 4 Comparison of the result of the proposed method with the related studies

	The proposed method	The method in the study [12]	The method in the study [5]
Models	Resnet 34	Resnet 18	Resnet 50
Training datasets	With **15,040** breast cancer X-ray images of the patients of Vietnam National Cancer Hospital	With **8,777** breast cancer X-ray images of the patients of the Vietnam National Cancer Hospital	With **7,912** breast cancer X-ray images of the patients of the Hanoi Medical University Hospital
Prediction hypotheses	macAUC = **0.84807** for the classification of Normal, Benign, and Malignant images	A macAUC = **0.82824** for the classification of Normal, Benign, and Malignant images	A macAUC = **0.754** for the classification of BI-RADS 045, BI-RADS 1, BI-RADS 23

We also observe that the accuracy of the proposed Resnet 18 model used 15,040 X-ray images for model training (which is the macAUC of 0.84807) for the classification of Normal, Benign, and Malignant images is higher than the accuracy of the Resnet 50 model used 7,912 X-Ray images for model training (which is the macAUC of the Resnet 50 model (which is the macAUC = 0.754) for classification of images into BI-RADS 045, BI-RADS 1, BI-RADS 23 categories. From our point of view, there are two reasons:

- In our study, the number of breast cancer X-ray images used for training the Resnet 34 model is higher than the number of breast cancer X-ray images used for training the Resnet 18 model in [12] and the Resnet 50 model in [5] that means the classification performance improves as the number of mammograms increases.
- The X-ray images as Benign, and Malignant images in our study used for model training are "smoother" than the X-ray images of BI-RADS 23, BI-RADS 045 categories i.e. the BI-RADS 23 included the images of BI-RADS 2, BI-RADS 3 and the BI-RADS 045 included the images of BI-RADS 0, BI-RADS 4, BI-RADS 5.

On the other side, Table 5 shows the comparison of the result of our model with the related study in [4].

In Table 5, we observe that the proposed Resnet 34 model used 15,040 breast cancer X-ray images collected by the Vietnam National Cancer Hospital for model training and achieves an AUC = **0.89758** for the classification of malignant/not malignant and an AUC = **0.78608** for the classification of benign/not benign which is higher than the accuracy of the 'view-wise' image-and-heatmaps ensemble modeling [4]. This comparison shows that the dataset for training the 'view-wise' image-and-heatmaps ensemble model is increased with over **1,000,000** images in comparison with **15,040** images of the Resnet 34 model.

Table 5 Comparison of the result of the proposed method with the related study for the classification of malignant/not malignant and benign/not benign

	The proposed method	The method in the study [4]
Models	Resnet 34	The 'view-wise' image-and-heatmaps ensemble model
Training datasets	With **15,040** breast cancer X-Ray images of the patients of Vietnam National Cancer Hospital	With over **1,000,000** high-resolution mammographic images
Prediction hypotheses	– an AUC = **0.89758** for the classification of malignant/not malignant – an AUC = **0.78608** for the classification of benign/not benign	– an AUC = **0.895** for the classification of malignant/not malignant – an AUC = **0.756** for the classification of benign/not benign

6 Conclusions

In this paper, we have presented the general CNN architecture of the classification system for classifying breast cancer X-ray images into three categories Normal, Benign, and Malignant. We have built a mammogram database consisting of 15,040 images of Normal, Benign, and Malignant classes used for model training. We used 15,040 breast cancer images for training the Resnet 18 to classify mammograms into three categories Normal, Benign, and Malignant. Our system achieves macAUC of 0.84807, an average sensitivity of 0.69058, and an average Specificity of 0.84336. We have compared the results of our study with the results of the studies in [12, 5], which shows that the results of our proposed method are more accurate than the results in [5, 12]. As of [1], "a supervised deep learning algorithm will generally achieve acceptable performance with around 5,000 labeled examples per category", our Resnet 34 model was trained by more than 15,000 breast cancer images for classifying three classes of Normal, Benign, and Malignant and achieves a macAUC of 0.84807. We also have compared the results of our study with the results of the studies in [4] which shows that our ResNet 34 receives an accuracy higher than the result in [4] used the 'view-wise' image-and-heatmaps ensemble model. This accuracy of the proposed model is acceptable because according to the reader study in [2], a committee of four human-experienced radiologists achieved the macAUC of 0.704, while our model achieved a macAUC of 0.84807 for the classification of three categories. In further works, we will continue to collect, as much as possible, breast cancer images for training CNN models and will apply some ensemble methods to increase the classification performance of the system.

Acknowledgements This work is partly supported by the Grant of State Project KC-4.0-18/19-25 "Research on the development of an intelligent system assisted in breast cancer diagnosis based on breast cancer X-ray images and pathology images". Thanks to the medical doctors of Vietnam

National Cancer Hospital for collecting and providing mammograms for this study. We thank Prof. Anh Nguyen from Auburn University, AL, USA, for his advice on this work.

References

1. Goodfellow, I., Bengio, Y., Courville, A.: Deep Learning. The MIT Press (2016)
2. Geras, K.J., Wolfson, S., Shen, Y., Wu, N., Kim, S.G., Kim, E., Heacock, L., Parikh, U., Moy, L., Cho, K.: High-Resolution Breast Cancer Screening with Multi-View Deep Convolutional Neural Networks (2018). arXiv: 1703.07047v3 [csCV]
3. Essam, H., Emam, M.M., Ali, A.A.: An optimized deep learning architecture for breast cancer diagnosis based on improved marine predators algorithm. Neural Comput. Appl. **34**, 18015–18033 (2022)
4. Wu, N., Phang, J., Park, J., Shen, Y., Huang, Z., Zorin, M., Jastrzebski, S., Févry, T., Katsnelson, J., Kim, E., Wolfson, S., Parikh, U., Gaddam, S., Lin, L.L.Y., Ho, K., Weinstein, J.D., Reig, B., Gao, Y., Toth, H., Pysarenko, K., Lewin, A., Lee, J., Airola, K., Mema, E., Chung, S., Hwang, E., Samreen, N., Kim, S.G., Heacock, L., Moy, L., Cho, K., Geras, K.J.: Deep neural networks improve radiologists' performance in breast cancer screening. IEEE Trans. Med. Imaging **39**, 1184–1194 (2019)
5. Hanh, B.M., Linh, L.T., Cuong, N.N., Nguyen, T.B., Doan, L.T., Le, C.D., Giao, V.T., Ly Ngo, T.L., Xuyen Hoang, T.H., Thang, N.D., Anh, N.T., Dan, N.D., Dung, N.V., Duc, T.V., Nguyen, Q.H., Nguyen, A., Phuong, N.H.: Convolutional neural networks improve radiologists' performance in breast cancer screening for vietnamese patients. Appl. Artif. Intell. **36**(1), e2151185 (3854–3872 pages) (2022)
6. Nitish, S., Geoffrey, H., Alex, K., Ilya, S., Ruslan, S.: Dropout: a simple way to prevent neural networks from overfitting. J. Mach. Learn. Res. **15**, 1929–1958 (2014)
7. "OpenCV: Histograms—2: Histogram Equalization.," 2022, https://docs.opencv.org/4.x/d5/daf/tutorial_py_hi-stogram_equalization.html. Last accessed October 2022
8. Cao, G., Huang, L., Tian, H., Huang, X., Wang, Y., Zhi, R.: Contrast enhancement of brightness-distorted images by improved adaptive gamma correction. Comput. Electr. Eng. **66**, 569–582 (2018)
9. Yoshua, B.: Learning deep architectures for AI. Found. Trends® Mach. Learn. **2**, 1–127 (2009)
10. GLOBOCAN 2020. Accessed at https://gco.iarc.fr/today/data/factsheets/populations/704-viet-nam-fact-sheets.pdf in September 2023
11. Phuong, N.H., Toan, H.M., Thi, N.V., Lam, N.L., Dung, N.K., Tu, D.V.: Segmentation of the abnormal regions in breast cancer X-Ray images using U-Net. In: Phuong, N.H., Kreinovich, N. (eds.) Deep Learning and Other Soft Computing Techniques: Biomedical and Related Applications, pp. 135–145. Springer, Berlin (2023)
12. Phuong, N.H., Toan, H.M., Tu, D.V., Khac-Dung, N., Thi, N.V., Lam, N.L., Nguyen, A.: Convolutional neural networks based classification of mammograms. In: Daimi, K., Al Sadoon, A. (eds.) Proceedings of the 2023 International Conference on Advances in Computing Research (ACR'23). ACR 2023. Lecture Notes in Networks and Systems, vol. 700, pp. 30–43. Springer, Cham (2023)

OAGRE: Outlier Attenuated Gradient Boosted Regression

John Hawkins

Abstract Gradient boosting is a general machine learning technique for iteratively building a model by fitting the residuals of an existing model. In spite of its genrality as a technique, there are several reasons why gradient boosting may fail to adequately model a given dataset. One of these reasons is the presence of excessive noise in the training data. In this paper we propose a technique that will mitigate the impact of noise under some conditions of heteroscedasticity. We break the assumption of independently distributed noise and derive a simple variation of the gradient boosting algorithm we name Outlier Attenuated Gradient Boosting Regression (OAGRE). We conduct experiments by generating large volumes of synthetic datasets with varying degrees of heteroscedasticity and benchmark our approach against multiple standard gradient boosting algorithms as well as simpler tree models. Our approach provides value in approx 30% of these synthetic datasets with an average improvement of between 7 and 12% lower root mean squared error. Based on these results we recommend the inclusion of the OAGRE method when testing algorithms on data with suspected heteroscedasticity issues.

1 Introduction

Gradient Boosting has proven itself to be one of the most significant advances in the field of machine learning in recent history. Its utility in providing accurate predictions for supervised learning is so strong that it has come dominate the field of competitive machine learning on tabular datasets (as explified by the competitions run by Kaggle [1]). In spite of its success, there remain a range of specific circumstances which limit the ability of gradient boosting (and other algorithms) to provide adequate reaults. One of these circumstances is the presence of complex noise in the data. Early boosting algorithms, like AdaBoost, were known to suffer in the presence of noise [2, 3]. Modern gradient boosting algorithms are less susceptible to these issues and tend to perform well with small amounts of heteroscedastic noise [4]. However,

J. Hawkins (✉)
Transitional AI Research Group & Getting-Data-Science-Done.com, Sydney, Australia
e-mail: john@getting-data-science-done.com

N. Hoang Phuong et al. (eds.), *Machine Learning and Other Soft Computing Techniques: Biomedical and Related Applications*, Studies in Systems, Decision and Control 543, https://doi.org/10.1007/978-3-031-63929-6_15

there is always potential for learning issues to emerge if some of the residuals become dominated by the presence of noise.

Gradient boosting algorithms are predeominantly tree based and there are multiple variations offering improvements in performance and comptational efficiency, typically through the way variables are selected and the base learners are combined [5]. However, in all cases the theoretical underpinnings of the gradient boosting algorithm remain unchanged. The algorithm consists of learning an ensemble of base learners as an additive combination. In each iteration the new model is trained to predict the residuals of the preceding ensemble [6–8].

In this paper we develop a variation of the gradient boosting algorithm by making several non-standard assumptions about the noise present in the data. This involves breaking the assumption that the noise is identically distributed across the problem space. In particular we assume that the distribution of the noise varies in a way that can be at least partially predicted by the known regressors for the machine learning problem.

Noisey data with hetereoscedasticity has been shown to be deleterious to regression trees (the most common base learner for gradient boosting models) [9]. In addition, there are many application domains in which the noise is known to be hetereoscedastic, and the focus of research is to develop custom algorithms that explicitly model the noise variance [10, 11], or build custom ensembles where subcomponents are used to accomodate the hetereoscedasticity [12]. Common general machine learning approaches include the development of noise filtering or re-labelling techniques, under the assumption that the noisy data is an outlier to be detected [13]. Other research has focused on modification of the base learners in ensemble methods (including gradient boosting ensembles) to make them more robust to hetereoscedasticity [14], or to attempt to explicitly model the heteroscedastic noise distribution [15, 16] and incorporate the noise distribution into the loss function to weight samples accordingly [17]. An alternative approach has been to augment the design of the gradient boosting method by quantifying the uncertainty in the predictions generated by the heteroscedasticity [4], or to model the output distribution directly using quantile or expectile regression [18]. In situations in which there is known noiseless data, or it can be simulated, then there are techniques that allow models to be trained such that they are influenced less by the noise distribution [19]. However, as our ability to fit models depends heavily on the amount of data and the nature of the noise, which is usually unknown, heteroscedasticity remains an open problem for machine learning research.

In this work we explore the extent to which assuming hetereoscedastic noise in a dataset can be used to derive a fundamental variation of the gradient boosting algorithm. The core idea is that if the noise distribution is variable then high noise regions of the problem space should be fit with simpler models than the low noise regions. Such that the model does not waste computational resources fitting noisy data points. The gradient boosting is selectively applied by predicting whether a given sample belongs to a region that surpasses the noise threshold for each level of the boosted model.

To do so we assume that there will be some correlation between the magnitude of the error made by the model and the presence of higher variance noise. This assumption requires only that our feature vector and model be rich enough that we expect the model to be able to learn the relationship between dependent and independent variables well enough that noisier samples stand out. In the absence of noise we expect error to be low, when noise is present we expect it to be high. Because the first assumption gives us non-uniform noise, we therefore expect the presence of a noise gradient that is correlated with the error.

Combining these two assumptions with the learning strategy of gradient boosting for regression, leads us to an alternative formulation of the gradient boosting algorithm. The core idea being that each iteration of boosting should seek to learn only from residuals that come from regions in the parameter space that contain less noise. In other words we predict which samples are noisier outliers and exclude them from the training of subsequent models in the ensembnle. We call this algorithm Outlier Attenuated Gradient Boosting Regression (OAGRE).

2 Method

We start the derivation of the method by assuming that the noise is not randomly distributed with respect to the features or predictors, but instead is predictably heteroscedastic. This assumption is contrary to the standard expression for the learning problem, an example of which is shown in Eq. 1 taken from Friedman 2000 [6].

$$y_i = F^*(X_i) + \epsilon_i \tag{1}$$

This expression defines the target variable of instance i, y_i as a function of the features vector X plus some Gaussian noise ϵ_i. It is generally assumed that the noise term is independent of the feature vector X. It may correspond to inherent noise, or be due to unobserved causal factores that are not present in the features, and hence limit our ability to perfectly represent the target using the given features.

We modify this assumption in the following way: the noise term ϵ_i is partially dependent upon the feature vector X_i, but only in its variance. In real world processes there are multiple reasons why noise might be correlated with predictors. Noise can come from sampling problems, which are not uniformly distributed. Noise might be due to limitations of the devices or methodology used to collect data, which again tend not to be uniformly distributed. Most noise is due to some physical process, which itself is regulated by variables, if any of the variables that regulate the noise are present as features then we should expect the statistical structure of the noise to vary over the features space.

This means that we should expect the learning problem in many real world problems to be

$$y_i = F^*(X_i) + \epsilon(N(X_i)) \tag{2}$$

in which $\epsilon(N(X_i))$ refers to a sample drawn from a noise distribution N that is dependent on the independent variables of the sample X_i.

When we are training a model to predict y_i, then we typically depict the fit of the model using a similar equation.

$$\hat{y}_i = \hat{F}(X_i) \tag{3}$$

Such that the function $\hat{F}$ has been fit such that given a feature vector X_i it provides an estimate of y_i. The error of such a model can be calculated for each record as the difference between the actual and predicted value:

$$err_i = y_i - \hat{y}_i \tag{4}$$

Note that error is typically calculated using some loss function $L(y, \hat{y})$, however we will ignore that complication for the moment. We can substitute Eqs. 2 and 3 into Eq. 4 and with some rearrangement end up with the following expression for the error:

$$err_i = (F^*(X_i) - \hat{F}(X_i)) + \epsilon(N(X_i)) \tag{5}$$

In other words the expected error in our model is the difference between the true underlying function and the estimated one, plus the expected noise in that region of the problem space.

For many algorithms the change we have made to the distribution of noise is not something that can easily be exploited. However, the gradient boosting algorithm has the potential to exploit this new feature of the problem. Usually noise undermines a gradient boosting step because it is attempting to use the residual of the existing model to build a new iteration that improves the previous model. One might imagine that at some point the residuals come to be dominated by the noise term, and hence a gradient boosted algorithm cannot see the signal for the noise.

If however, the noise is not identically distributed across the feature space then we could potentially direct each round of gradient boosting to ignore noisy examples, and focus on learning from a subset of the data. This is similar to what Freund did in the development of BrownBoost [2], a generalisation of AdaBoost, although in his treatment it involved setting an expected minimum error rate and then smoothly downweighting samples that exceeded this threshold.

In order to modify the learning algorthm to ignore samples suspected of being dominated by noise we need to provide some kind of selection criteria for deciding at the start of each iteration which samples to exclude. We introduce an additional threshold meta-parameter which determines when an example should be ignored. This parameter determines the threshold in the standard deviation of the resisual errors beyond which a sample is considered an outlier. In addition we allow this parameter to be adjusted downward with each iteration of learning, so that the model increases the volume of samples that are treated as outliers as the boosting ensemble grows.

In order to apply the model to new data we cannot simply make predictions using all stages of the gradient boosting because deeper level trees have been trained only samples from a subset of the feature space. In other words, our gradient boosted model covers the problem space at different levels of granularity, depending on our expectations of the noise distribution. We only apply each subsequent gradient boosting model in a way that it is attenuated according to our estimate of whether or not it is appropriate. We do this by fitting a classification model at each stage of gradient boosting which predicts whether or not a given example belongs to the signal set (of relatively low noise examples). We use the classifier to build a model to estimate the probability that sample i is considered to be low noise at depth j in our gradient boosted model: $p_j(lownoise|X_i)$.

The final prediction can be calculated according to the following formula:

$$\hat{y}_i = \hat{B}(X_i) - \sum_{j=1}^{M} p_j(lownoise|X_i) * \hat{R}_j(X_i) \tag{6}$$

where M is the parameter defining the number of models in the ensemble, $\hat{B}$ is the base regression model fitted to the entire dataset, and $\hat{R}_j$ is the estimate of the residuals at depth j of the boosting. Our prediction $\hat{y}_i$ is a combination of the base model with the negative sum of all residual estimates over the M levels of boosting each of which is attenuated by the estimate of the probability that the sample is a low noise example at that depth of the boosted model.

2.1 Implementation

```python
from sklearn.tree import DecisionTreeClassifier
from sklearn.tree import DecisionTreeRegressor
from oagre import OAGRE

oagre = OAGRE(
    classifier=DecisionTreeClassifier(max_depth=5),
    regressor=DecisionTreeRegressor(max_depth=5)
)

oagre.fit(X_train, y_train)
...
```

Listing 1 Usage of OAGRE with scikit-learn models

We have implemented the OAGRE method as a meta-learning algorithm compatible with scikit-learn [20], and made it available as an open source python library in the GitHub repository (https://github.com/john-hawkins/oagre). The implementation requires that the user provide two base models, a classifier and a regressor, which it then uses to building the gradient boosted ensemble. We show an example of how to instatiate and train an OAGRE regressor in Listing 1, where we use the sklearn DecisionTree models for the base classifier and regressor.

Our implementation allows an OAGRE ensemble model to be included inside the machine learning framework scikit-learn pipelines [20]. The advantage of the pipeline approach is that an OAGRE model can be built into a single serialised model for deployment, including data preprocessing stages. The OAGRE library can be installed as a package from the PyPi repository (https://pypi.org/project/oagre) or installed directly from the source code (https://github.com/john-hawkins/oagre).

2.2 Evaluation

To evaluate the effectiveness of our approach we conduct an experiment in which large volumes of synthetic data sets are created with varying parameterisation of the hetereoscedasticity in the error term. The parameters of the simulation are shown in Table 1. We generate 100 random datasets for each combination of parameters, such that the independent variables are drawn from a range of different numercial distributions and the target variable is created through a randomly generated formula combining linear and non-linear combination of a random subset of independent variables. We generate a second random function that is fed through a logistic function to create a probability distribution. The output of this probability distribution is used to determine which datapoints will be subject to excessive noise, thus ensuring that the data is not onluy hetereoscedastic, but that this property is partially dependent on the independent variables.

Once the data is generated we take a random split of 90% for training and 10% for testing. We train six models for evaluation and comparision. The first two are standard implementations of the Gradient Boosted Tree algorithm, the GBM model built into the scikit-learn package, and the LightGBM model [5]. For each of these models we train them using default parameters, in order to reduce the computational complexity of performing hyperparameter tuning over large volumes of synthetic data. The second two models are OAGRE ensembles using different base learners. The first is trained with an ExtraTrees model (a Random Forest Implementation) and

Table 1 Simulation parameters

Parameter	Description	Values
Dataset size	Number of records in the dataset	2000, 5000, 10000
Variables	Number of independent variables	15, 20, 25, 30
Outlier proportion	Proportion of records with larger noise	0.4, 0.5, 0.6, 0.7
Std multiplier	The factor applied to noise std of outliers	Random int (3,8)

Table 2 Models and initialisation parameters

Model	Description	Estimators	Max depth
GBM	SKLearn GB regressor	100	3
LGBM	LightGBM	100	Unbounded
OARGE-DT	OAGRE with DecisionTrees	20	5
OAGRE-XT	OAGRE with extratrees	200	5
XTR	ExtraTrees regression	10	5
DTR	DecisionTree regression	1	5

the second with a DecisionTree model, both of these are taken from the standard scikit learn implementation. Finally, we included the base regression models used by the two OAGRE models for comparison, to ensure that the OAGRE algorithm was adding value to these models. The standard GBM models were initialised with default paramaters, the other models were initialised shown in Table 2. These initialistions are based on our experience with applying the OAGRE technique. Note, that when using the ExtraTrees model as a base learner, the number of estimators is a combination of the estimators used by ExtraTrees and the estimators used in the boosting layers of OAGRE.

The complete code for the simulation and evaluation is provided in our GitHub repository. We note that as the generation of random formulas sometimes produced null values, there are significant checks in place to regenerate data when problematic scenarios occur. In total this process produced 2, 185 datasets that the models were trained and evaluated on.

3 Results

Table 3 shows the proportion of datasets for which each model provided the best result (measured in terms of lowest Root Mean Squared Error). In addition, we show the average percentage improvement of each model against the next best model for the task. As we see, the overall winner is the standard Gradient Boosting Machine implementation from the scikit-learn library. The two OAGRE implementations provide a superior model in approx 30% of datasets, with the version using ExtraTrees models as the base learner showing stronger performance, both in terms of the proportion of times it is superior, and the expected percentage of improvement. The two base models alone are only superior in a small percentage of instances, with a more marginal improvement.

We took the parameters that generated these synthetic datasets and analysed multiple factors that might contribute to the circumstances in which the OAGRE approach

Table 3 Overall results

Model	Total Wins	Percentage of Wins (%)	Average improvement (%)
GBM	890	40.7	30.8
LGBM	274	12.5	25.4
OARGE-DT	279	12.7	11.6
OAGRE-XT	430	19.7	7.3
XTR	131	6.0	2.5
DTR	180	8.2	2.4

added value. This included the outlier proportion and standard deviation multiplier factor used to generate the data, as well as the proportion of non-linear elements used in the random formula that generated the target variable. We observed that the proportion of datasets upon which OAGRE provided superior performance appeared to be increasing with the proportion of the generating function that was non-linear, as shown in Fig. 1.

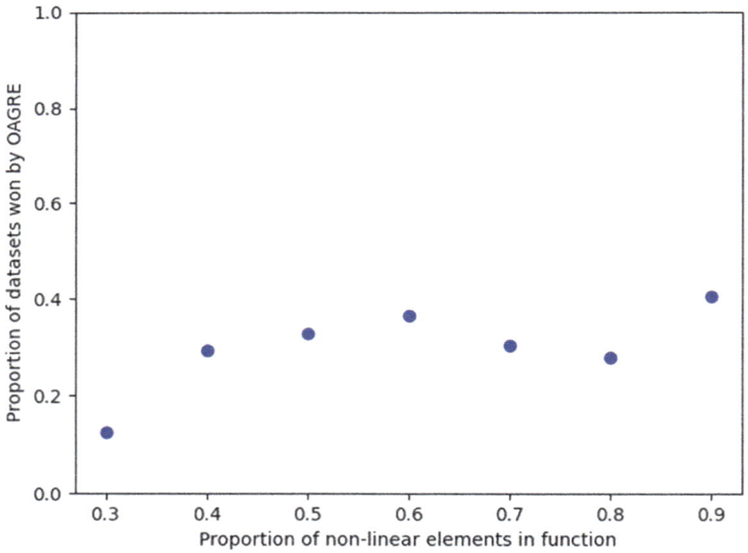

Fig. 1 OAGRE Wins against Non-linear Elements in Target Function

4 Conclusion

By questioning a standard assumption in the definition of regression problems, pertaining to the nature of the noise present in a data set, we have been able to derive a variation of the gradient boosting algorithm. Our variation compensates for the presences of heteroscedasticity by explcitly modelling the probability that a data point comes from a high noise region of the data space, and then excluding it from the residual estimates in the boosting process. We incorporate the probability of a data point being a noisy outlier at prediction time to attentuate the impact of residual estimates in the boosting process. The resulting meta-learning approach to building an ensemble we call Outlier Attenuated Gradient Boosting Regression and we release it as a scikit-learn compatible open source library.

We conducted a simulation based experiment in which we generated large volumes of datasets with hetereoscedastic noise imposed through parameteristation. These experiments indicate that our approach will provide superior performance to standard gradient boosting implementation in approx 30% of cases, with an average reduction in root mean squared error of between 7% and 12% (compared to the next best model). We note that this work was done using default parameters to limit the complexity of the search space, hence the impact of significant parameter tuning is unknown, but would likely to shift some of these results. Nevertheless, we observe that the proportion of instances in which the OAGRE approach proved superior was approximately equal to its proportion in the set of methods tested. Hence, we conclude that the technique will likely provide value for some regression problems with heteroscedastic noise. For this reason we recommend that practitioners facing a regression problem with noiesy heteroscedastic data include the OAGRE algorithm within a set of empirically evaluated methods.

Future research could be conducted to determine the exact properties of a dataset that make it amenable to our technique. However, preliminary analysis suggests that the more non-linear the function that is being estimated, the greater the likelihood that OAGRE will provide predictive value.

References

1. Fogg, A.: Anthony Goldbloom gives you the secret to winning Kaggle competitions. Blogpost (2016). https://www.import.io/post/how-to-win-a-kaggle-competition/
2. Freund, Y.: An adaptive version of the boost by majority algorithm. Mach. Learn. **43**(3), 293–318 (2001)
3. Schapire, R.E.: The boosting approach to machine learning: an overview. In: Nonlinear Estimation and Classification, pp. 149–171. Springer, New York, NY (2003)
4. Brophy, J., Lowd, D.: Instance-based uncertainty estimation for gradient-boosted regression trees. In: Proceedings of the 36th Conference on Neural Information Processing Systems (NeurIPS 2022) (2022)

5. Ke, G., Meng, Q., Finley, T., Wang, T., Chen, W., Ma, W., Ye, Q., Liu, T.-Y.: LightGBM: a highly efficient gradient boosting decision tree. In: Proceedings of the 31st International Conference on Neural Information Processing Systems, NIPS'17, (Red Hook, NY, USA), pp. 3149–3157. Curran Associates Inc. (2017)
6. Friedman, J., Hastie, T., Tibshirani, R.: Additive logistic regression: a statistical view of boosting. Ann. Stat. **28**, 337–407 (2000)
7. Friedman, J.: Greedy function approximation: a gradient boosting machine. Ann. Stat. **29**(5), 1189–1232 (2001)
8. Friedman, J.: Stochastic gradient boosting. Comput. Stat. Data Anal. **38**, 367–378 (2002)
9. Ruth, W., Loughin, T.: The Effect of Heteroscedasticity on Regression Trees (2016)
10. Zhenxing, L., Xiaodan, H., Hao, K., Chen, L., Huang, B.: Gaussian process regression with heteroscedastic noises—a machine-learning predictive variance approach. Chem. Eng. Res. Des. **157** (2020)
11. Zhang, J., Guo, F., Hao, K., Huang, B., Chen, L.: Identification of errors-in-variable system with heteroscedastic noise and partially known input using variational bayesian. IEEE Trans. Ind. Inf. 19(10), 10014–10023 (2023)
12. Zeng, H., Shao, B., Bian, G., Dai, H., Zhou, F.: A hybrid deep learning approach by integrating extreme gradient boosting-long short-term memory with generalized autoregressive conditional heteroscedasticity family models for natural gas load volatility prediction. Energy Sci. Eng. **10** (2022)
13. Ustinovskiy, Y., Fedorova, V., Gusev, G., Serdyukov, P.: Meta-gradient boosted decision tree model for weight and target learning. In: Proceedings of the 33rd International Conference on International Conference on Machine Learning—Volume 48, ICML'16, pp. 2692–2701 (2016). JMLR.org
14. Henrey, A.J.: Statistical Learning Tools for Heteroskedastic Data. Ph.D. thesis, Simon Fraser University (2016)
15. Natarajan, R.: Gradient boosting for joint regression modeling of mean and dispersion. In: Research Report W0909–085. IBM Research Division, New York (2009)
16. Thomas, J., Mayr, A., Bischl, B., Schmid, M., Smith, A., Hofner, B.: Gradient boosting for distributional regression: faster tuning and improved variable selection via noncyclical updates. Stat. Comput. **28**, 1–15 (2018)
17. Lee, G.-H., Yang, S.-W., Lin, T.-H., Lin, S.-D.: Heteroscedastic learning and its realization on matrix factorization and gradient boosting regression/classification. Technical report (2017)
18. Yang, Y., Zou, H.: Nonparametric multiple expectile regression via er-boost. J. Stat. Comput. Simul. **85**, 1–17 (2014)
19. Wu, Z., Luo, J., Rincon, D., Christofides, P.: Machine learning-based predictive control using noisy data: Evaluating performance and robustness via a large-scale process simulator. Chem. Eng. Res. Des. **168** (2021)
20. Pedregosa, F., Varoquaux, G., Gramfort, A., Michel, V., Thirion, B., Grisel, O., Blondel, M., Prettenhofer, P., Weiss, R., Dubourg, V., et al.: Scikit-learn: machine learning in python. J. Mach. Learn. Res. **12**(85), 2825–2830 (2011)

Improve the Effectiveness of Predicting Student Dropouts Based on Deep Learning and SMOTE Models

Xuan Tho Dang

Abstract In the current era, the number of educational institutions is rapidly increasing, making the admissions process more challenging. Students have numerous options when choosing an educational institution to attend. Therefore, transferring schools or dropping out has become a noteworthy issue in educational institutions. Predicting student attrition and transfers holds significant importance in enhancing the quality of education and improving the teaching and learning experience. It helps educational institutions identify the core reasons leading to student attrition or transfers and provides timely and effective solutions. Predicting student performance is often a challenging task due to various factors, with one significant factor being the imbalance in real-world data collection. This article compares different data balancing techniques such as Random Over Sampling, SMOTE, and ADASYN for data preprocessing and balancing. Additionally, the article compares the effectiveness of traditional classification models like Random Forest, Support Vector Machine (SVM), K Nearest Neighbors (KNN), Naïve Bayes, Decision Tree, AdaBoost, and Gradient Boosting with newer approaches involving Deep Learning using various architectures to handle imbalanced datasets. Experimental results indicate that the combination of data balancing algorithms such as SMOTE and ADASYN with Deep Learning models yields the best predictive outcomes.

Keywords Deep learning · Education data · SMOTE · Student dropout prediction · Machine learning

X. T. Dang (✉)
Academy of Policy and Development, Ha Noi, Viet Nam
e-mail: thodx@apd.edu.vn

N. Hoang Phuong et al. (eds.), *Machine Learning and Other Soft Computing Techniques: Biomedical and Related Applications*, Studies in Systems, Decision and Control 543, https://doi.org/10.1007/978-3-031-63929-6_16

1 Introduction

The issue of predicting students' academic performance has garnered significant attention from researchers. Student academic performance refers to their mastery of knowledge and skills in the subjects they study, either individually or in collaboration with a group to solve specific tasks [1]. This, in turn, enables educational institutions to identify the fundamental reasons leading to student attrition or transfers, allowing for timely and effective solutions. In the broader context, it contributes to enhancing the overall quality of education in the country and improving the efficiency and effectiveness of educational institutions. This is a matter of concern not limited to any particular level of education or regional locality; it plays a crucial role in a nation's educational system as a whole [2–4].

Educational data mining is the process of predicting the academic outcomes of students based on data collected from their educational experiences. It holds significant importance in improving the quality of education and enhancing the teaching and learning processes. Educational data mining involves the collection, extraction, analysis, and utilization of data related to educational systems to understand and improve the teaching and learning processes. This entails using various data analysis methods and tools to uncover trends, models, and valuable insights from different data sources within the field of education [5]. The objectives of educational data mining include:

(a) Assessing Learning Performance: Analyzing data regarding students' academic outcomes to evaluate their performance. This can help identify specific issues and propose improvement measures.
(b) Predicting and Monitoring Learning Progress: Using data to predict the learning progress of students and provide information for tracking their development. This can assist teachers and administrators in making informed decisions to support learning.
(c) Optimizing Teaching: Analyzing data on how teachers teach and how students learn to identify the most effective teaching methods. This can help create a better learning environment for students.
(d) Detecting Trends and Issues: Using data to detect trends in the teaching and learning processes, as well as identifying specific challenges that students and teachers may be facing.
(e) Proposing Curriculum Enhancements: Based on data on learning performance and feedback from students, proposing enhancements to curricula to create a better learning experience for learners.

According to an independent report by the European Commission, there is a significant dropout rate among students before completing higher education courses [2]. Even in the most successful countries, such as Denmark, only around 80% of students complete their programs, while in Italy, this figure drops to 46%. The report highlights key factors contributing to student dropout, with socioeconomic conditions being the primary cause. In recent years, early prediction of student outcomes has garnered increasing research interest [3].

In this study, we focus on constructing a machine learning model to better predict students at risk of dropping out, thereby enabling educational institutions and administrators to implement more suitable policies to support these students. Experimenting with a standard dataset containing 4,424 records, we identified an issue of data imbalance, which resulted in machine learning models achieving suboptimal accuracy and reduced predictive performance [6]. We applied several data preprocessing methods to address this imbalance issue and employed a range of state-of-the-art machine learning algorithms. Some key contributions of this paper include:

(a) Analysis of the challenges faced by standard machine learning algorithms in predicting student dropout when the number of dropout cases is a minority compared to non-dropout cases.
(b) Evaluation of the results when applying data balancing algorithms to the student dropout prediction problem.
(c) Assessment of the outcomes when applying machine learning algorithms to both the original and balanced datasets. Particularly, deep learning algorithms demonstrated superior results compared to other methods.

The following sections of this article are organized as follows: Sect. 2 provides a literature review in this field, assessing the advantages and disadvantages of previous methods. Our proposed approach is described in Sect. 3, elucidating its components and basic strategy. Section 4 presents detailed test results and analyses, shedding light on the effectiveness and performance of our method. Finally, Sect. 5 summarizes our findings, draws conclusions based on the results, and considers potential directions for future research and advancement.

2 Related Work

Data mining in education has garnered significant attention from researchers. This is because educational data holds great potential for providing valuable insights into the learning process of students, the performance of educators, and the effectiveness of educational programs [7–15]. From these data sources, data scientists and experts can develop better teaching and learning methods, customize learning experiences for individual students, and even predict future student success or dropout.

This study of Mengash [7] explores ways to enhance university admissions decisions using data mining techniques for predicting applicants' academic success. It validates the proposed methodology using a dataset of 2,039 students enrolled at a Saudi public university's Computer Science and Information College from 2016 to 2019. The findings reveal that early university performance can be forecasted before admission based on specific pre-admission criteria, including high school grades, Scholastic Achievement Admission Test scores, and General Aptitude Test scores. Notably, the Scholastic Achievement Admission Test score is the most accurate predictor, suggesting it should carry more weight in admissions systems. Furthermore, the study identifies the Artificial Neural Network as the superior technique with an accuracy rate exceeding 79% compared to other considered methods like Decision Trees, Support Vector Machines, and Naïve Bayes.

Francis [8] introduces a novel prediction algorithm for assessing student academic performance, utilizing both classification and clustering techniques. The algorithm is tested in real-time using student datasets from diverse academic disciplines in Kerala, India. The findings demonstrate that the hybrid algorithm, which integrates clustering and classification approaches, outperforms others in accurately predicting student academic performance.

Waheed [9] employs a deep artificial neural network and handcrafted features derived from virtual learning environments' clickstream data to predict students at risk and enable early intervention measures. The model achieves a classification accuracy ranging from 84 to 93%, surpassing logistic regression (79.82–85.60%) and support vector machine models (79.95–89.14%). Including legacy and assessment-related data significantly impacts the model's performance, and students engaging with previous lecture content exhibit improved outcomes. This study aims to assist institutions in developing a pedagogical support framework and enhancing higher education decision-making for sustainable education.

In Sekeroglu's paper [10], two datasets are utilized to predict and classify student performance with the assistance of five machine learning algorithms. The study involves eighteen experiments, and initial findings indicate the potential predictability of student performance and the enhancement of classification through data preprocessing techniques prior to applying machine learning algorithms.

Lorenz Kemper and colleagues in their research [11] have proposed a method to successfully predict dropout incidents in education based on the academic records of individual students. Given that, nowadays, such data is digitally recorded in most higher education institutions, their similar approaches could be implemented in other institutions without incurring significant initial costs. The authors employed logistic regression and decision trees, yielding promising results in terms of prediction accuracy. However, they also analyzed the logistic regression model's drawback of complexity in handling more intricate data preparations. Furthermore, this research also highlighted an imbalance issue in the data, where the number of dropout cases was less than other student categories in the original dataset. This represents a potential area for further development to enhance predictive effectiveness.

Based on the analyses above, in our paper, we propose addressing some of these issues to enhance the accuracy in predicting student dropout. Specifically:

(1) Updating and supplementing additional information beyond the students' subject grades. These attributes, to some extent, will also influence the decision to drop out.
(2) Implementing data balancing methods will help address the challenge of predicting student dropout with limited instances in the dataset.
(3) Applying advanced deep learning models will enhance the prediction.

3 The Method

In this research, we constructed and evaluated machine learning models to predict academic outcomes and dropout using a standardized dataset provided by the Polytechnic Institute of Portalegre [15]. Figure 1 illustrates the workflow of our approach, comprising five steps: (1) Employing k-fold cross-validation to split the original dataset into training and testing datasets; (2) Applying data balancing algorithms such as ROS, SMOTE, and ADASYN to address data imbalance; (3) Implementing machine learning algorithms including RF, SVM, Decision Tree, Adaboost, Gradient Boosting, and Deep Learning to build models; (4) Evaluating the performance of these models.

3.1 Datasets

The dataset utilized in this paper contains data sourced from a university-level educational institution, encompassing various attributes related to university students, including demographics, socio-economic factors, and academic outcomes, to investigate the impact of these factors on student dropout or graduation [15]. This dataset comprises 35 attribute columns and 4424 data instances. The data attributes in the first 34 columns contain personal information of the students, which has been encoded into numerical values. Column 35, denoted as "Target" represents the academic performance or outcome of the students. The "Target" attribute is divided into three

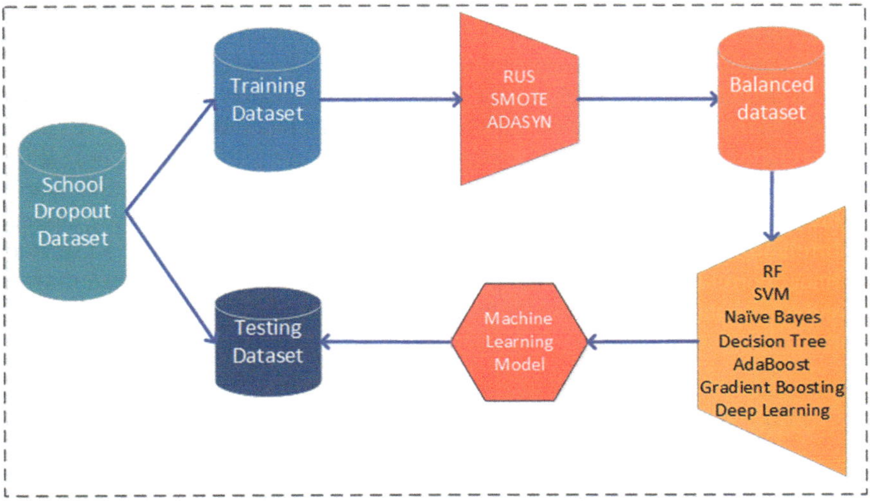

Fig. 1 The workflow consists of four fundamental steps

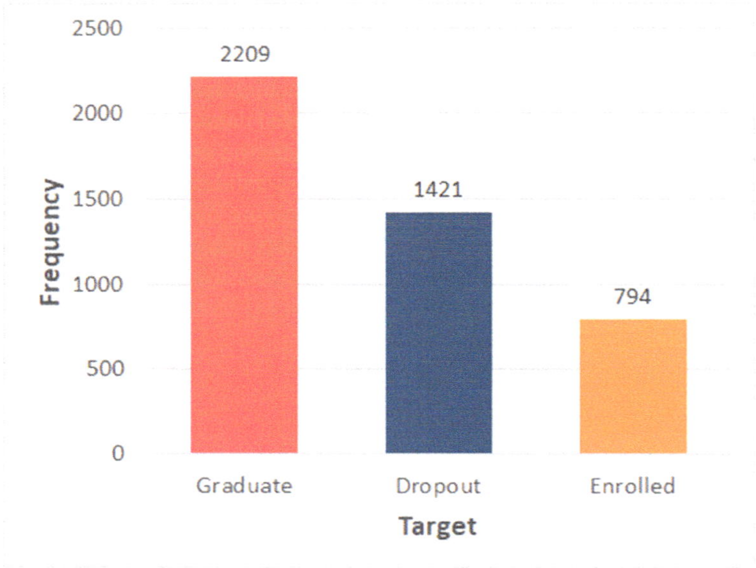

Fig. 2 The allocation of student records across the three categories evaluated for academic achievement

class labels with proportions as depicted in Fig. 2 below. Detailed information about the student data attributes is presented in Table 1.

The correlation relationships between the attributes in the educational dataset under consideration are visualized in a Heatmap, represented as a color-coded matrix. The values on the Heatmap indicate the degree of correlation between the attributes, with the colors providing additional information about the strength of these correlations, as depicted in Fig. 3 below. The results of a correlation analysis can provide valuable insights into the relationships between two attributes. Based on the results of the correlation analysis, it can be concluded that the current attributes do not exhibit a strong correlation with the Target attribute. This finding can have various implications, but some key points to consider are: No correlation doesn't imply absence of correlation: Although there is no strong correlation, there may still exist weak or nonlinear correlations between the attributes. This can be further explored through more detailed statistical analysis or by employing different methods to uncover the relationships between them.

3.2 Data Imbalance Preprocessing

Data preprocessing in machine learning is an important and necessary step to prepare input data before feeding it into a machine learning model. In particular, datasets collected in real-world scenarios often exhibit data imbalance, which can lead to several

Table 1 List of attributes in the data

No	Attribute	Type	No	Attribute	Type
1	Marital status	Numeric/discrete	18	Application mode	Numeric/discrete
2	Nationality	Numeric/discrete	19	Application order	Numeric/ordinal
3	Displaced	Numeric/binary	20	Course	Numeric/discrete
4	Gender	Numeric/binary	21	Daytime/evening attendance	Numeric/binary
5	Age at enrollment	Numeric/discrete	22	Previous qualification	Numeric/discrete
6	International	Numeric/binary	23	Curricular units 1st sem (credited)	Numeric/discrete
7	Mother's qualification	Numeric/discrete	24	Curricular units 1st sem (enrolled)	Numeric/discrete
8	Father's qualification	Numeric/discrete	25	Curricular units 1st sem (evaluations)	Numeric/discrete
9	Mother's occupation	Numeric/discrete	26	Curricular units 1st sem (approved)	Numeric/discrete
10	Father's occupation	Numeric/discrete	27	Curricular units 1st sem (grade)	Numeric/continuous
11	Educational special needs	Numeric/binary	28	Curricular units 1st sem (without evaluations)	Numeric/discrete
12	Debtor	Numeric/binary	29	Curricular units 2nd sem (credited)	Numeric/discrete
13	Tuition fees up to date	Numeric/binary	30	Curricular units 2nd sem (enrolled)	Numeric/discrete
14	Scholarship holder	Numeric/binary	31	Curricular units 2nd sem (evaluations)	Numeric/discrete
15	Unemployment rate	Numeric/continuous	32	Curricular units 2nd sem (approved)	Numeric/discrete
16	Inflation rate	Numeric/continuous	33	Curricular units 2nd sem (grade)	Numeric/continuous
17	GDP	Numeric/continuous	34	Curricular units 2nd sem (without evaluations)	Numeric/discrete
			35	Target	Categorical

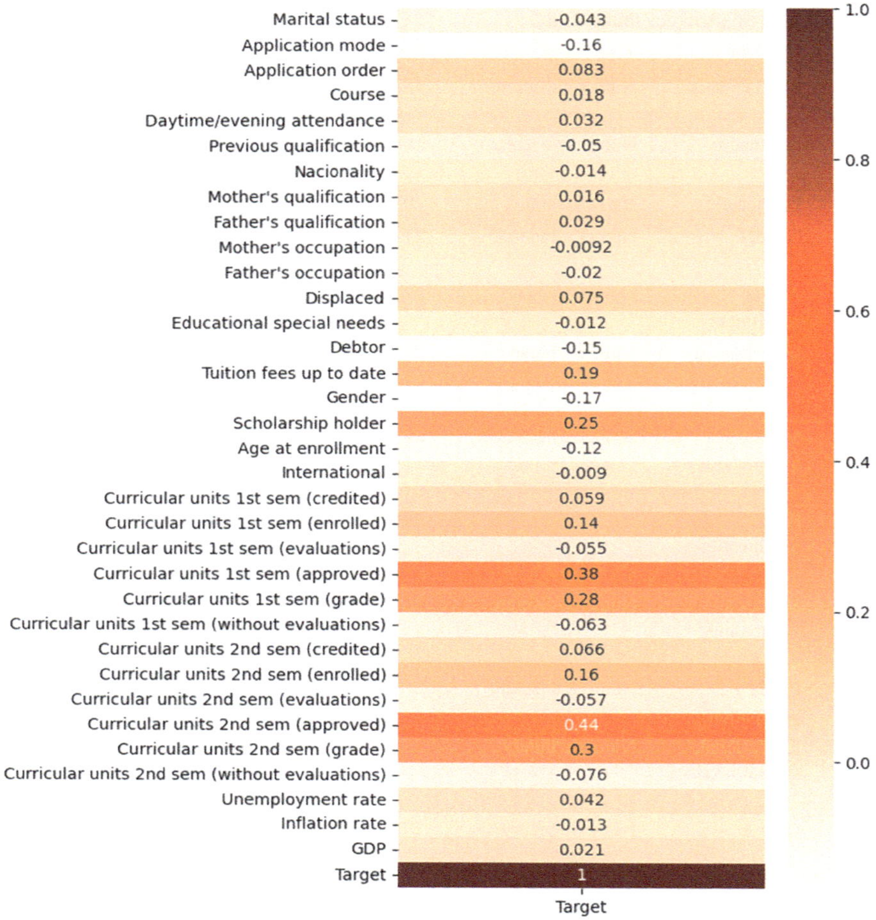

Fig. 3 The correlation between the attributes and the Target attribute

challenges in analysis and prediction. Some key issues related to data imbalance are as follows: (a) Ineffective Models: When dealing with imbalanced data, machine learning models tend to become ineffective. This can result in inaccurate predictions or bias towards the majority class if the imbalance issue is not addressed. (b) Impact on Accuracy: If one of the classes in imbalanced data has a small number of observations, the model may tend to predict the majority class without considering the minority class. This can lead to accuracy measurements that do not reflect reality and can skew the model's performance evaluation. (c) Increased Bias and Sensitivity: Models can become biased and sensitive to minority samples in imbalanced data. This means that a small change in the data can lead to significant variations in the model's predictions, causing instability and unreliability. Therefore, it is evident that data imbalance is a significant issue in data analysis and machine learning and needs to be carefully addressed to ensure accurate performance evaluation and predictions in real-world applications.

The dataset used in the paper exhibits data imbalance, as shown in Fig. 2, where the percentage distribution of class labels in the dataset is uneven. In this study, we investigate and apply several data preprocessing algorithms to address this data imbalance issue. Some of the data balancing algorithms introduced are as follows:

(a) Random Over Sampling (ROS) [16] is a data balancing method in machine learning, aims to mitigate this problem by increasing the number of instances in the minority class, thereby creating a more balanced distribution of classes in the dataset. This technique randomly selects and duplicates samples from the minority class until the class distribution becomes more equitable. By doing so, it helps machine learning models to better learn and generalize patterns from both classes, ultimately improving their predictive accuracy, especially for the minority class.

(b) SMOTE (Synthetic Minority Over-sampling Technique) [17] is a popular and well-known algorithm for handling imbalanced data, especially in classification tasks. This algorithm was developed to address the problem when the minority class in the dataset has a low proportion, causing bias and inaccuracy in machine learning models. SMOTE works by generating synthetic data samples for the minority class. It does this by randomly selecting a sample from the minority class and creating new samples by combining it with its nearest neighbors. This process helps augment the data for the minority class, improve class balance, and reduce model bias. SMOTE has become a crucial tool in addressing data imbalance and has been widely used in real-world applications to enhance the performance of machine learning models for classification and prediction [10].

(c) ADASYN (Adaptive Synthetic Sampling) [18] is a significant algorithm in dealing with imbalanced data. Similar to SMOTE, ADASYN also focuses on generating synthetic samples for the minority class to balance the class distribution in the dataset. ADASYN's strength lies in its adaptability to the data. ADASYN uses a distance-based method to determine the level of data imbalance and generates more synthetic samples for instances that are more minority. This intelligently enhances the minority class, ensuring that balance is achieved more effectively. ADASYN has been proven to be effective in many real-world applications and is an important tool in handling imbalanced data to improve the performance of machine learning models in classification and prediction scenarios.

4 Experiments

There are various machine learning models used for different classification tasks. This paper compares the effectiveness of several models, including Random Forest [19], Support Vector Machine (SVM) [20], K-Nearest Neighbors (KNN) [21], Naïve Bayes [22], Decision Tree [23], AdaBoost [24], Gradient Boosting [25], and Deep Learning [26]. The preprocessed data is split into two sets: the training dataset and the

Table 2 The prediction results on the original dataset without applying preprocessing algorithms

Model	Recall	Precision	F1-score	Accuracy
Random forest	**70.25**	**76.29**	**71.7**2	**78.76**
Support vector machine	66.88	71.76	67.78	76.95
KNN	56.92	63.43	56.68	69.27
Naive Bayes	60.65	63.34	61.21	69.72
Decision tree	65.00	64.89	64.89	70.28
AdaBoost	66.86	71.20	67.67	76.16
Gradient boosting	70.16	74.14	71.33	78.19
Deep learning	63.12	63.42	63.23	70.17

Table 3 The prediction results on the dataset using the Random Over Sampling method

Model	Recall	Precision	F1-score	Accuracy
Random forest	**71.18**	**72.43**	**71.24**	**71.07**
Support vector machine	65.91	67.50	66.24	65.83
KNN	57.05	58.57	56.76	56.60
Naive Bayes	53.32	53.59	50.13	52.20
Decision tree	55.85	55.78	55.79	55.56
AdaBoost	67.25	67.72	66.89	66.88
Gradient boosting	70.79	71.49	70.83	70.65
Deep learning	60.05	59.99	59.98	59.75

Table 4 The prediction results on the dataset using SMOTE method

Model	Recall	Precision	F1-score	Accuracy
Random forest	84.98	85.17	84.92	84.92
Support vector machine	78.14	78.97	78.16	78.05
KNN	77.29	78.14	77.37	77.38
Naive Bayes	65.17	65.33	65.18	65.08
Decision tree	73.44	73.67	73.48	73.45
AdaBoost	74.02	74.16	73.98	73.91
Gradient boosting	80.28	80.46	80.15	80.17
Deep learning	**86.60**	**86.76**	**86.6**3	**86.65**

testing dataset, with a 10-fold cross-validation. The obtained results are as follows (Table 2, 3, 4 and 5):

According to the results, we can observe that applying Deep Learning yields the best experimental F1-Score values among all classification methods, with values

Table 5 The prediction results on the dataset using ADASYN method

Model	Recall	Precision	F1-score	Accuracy
Random forest	83.19	83.64	83.26	83.27
Support vector machine	76.82	78.33	77.08	77.08
KNN	75.25	77.19	75.52	75.83
Naive Bayes	59.03	59.00	58.12	58.36
Decision tree	74.94	74.99	74.95	74.87
AdaBoost	71.78	71.62	71.64	71.63
Gradient boosting	78.13	78.64	78.13	78.19
Deep learning	**85.87**	**86.26**	**85.98**	**86.07**

exceeding 86%, especially when employing data balancing methods such as SMOTE. This demonstrates that the proposed method is robust and superior to the previous approaches.

5 Conclusions

This article proposes a dropout prediction method using deep learning techniques to leverage the database within the student management system at the university. After collecting the data, we conducted attribute selection, data preprocessing, and designed various classification methods, including the proposed Deep Learning combined with data balancing methods like ROS, SMOTE, and ADASYN. Experimental results demonstrate that the proposed model provides highly accurate prediction results and is entirely feasible for real-world applications. Further improvements to this model can be made by adding additional input attributes, such as incorporating more course grade attributes from semesters 1 and 2, and fine-tuning model parameters to enhance prediction performance. Additional comparisons with other advanced techniques should also be considered for future research.

References

1. Behr, A., Giese, M., Teguim Kamdjou, H.D., Theune, K.: Motives for dropping out from higher education-an analysis of Bachelor's Degree Students In Germany. European J. Edu. **56**(2), 325–343 (2021)
2. Kehm, B.M., Larsen, M.R., Sommersel, H.B.: Student dropout from Universities in Europe: a review of Empirical Literature. Hungarian Edu. Res. J. **9**(2), 147–164 (2019)
3. Namoun, A., Alshanqiti, A.: Predicting student performance using data mining and learning analytics techniques: a systematic literature review. Appl. Sci. **11**(1), 237 (2020)

4. Saa, A. A., Al-Emran, M., Shaalan, K.: Mining student information system records to predict students' academic performance. In: The International Conference On Advanced Machine Learning Technologies and Applications (AMLTA2019), vol. 4, pp. 229–239. Springer International Publishing (2020)
5. Martins, M.V., Tolledo, D., Machado, J., Baptista, L.M., Realinho, V.: Early prediction of student's performance in higher education: a case study. In: Trends and Applications on Information Systems and Technologies, vol. 1–9, pp. 166-175. Springer International Publishing (2021)
6. Ali, H., Salleh, M.N.M., Saedudin, R., Hussain, K., Mushtaq, M.F.: Imbalance class problems in data mining: a review. Indonesian J. Electr. Eng. Comput. Sci. **14**(3), 1560–1571 (2019)
7. Mengash, H.A.: Using data mining techniques to predict student performance to support decision making in University Admission Systems. IEEE Access **8**, 55462–55470 (2020)
8. Francis, B.K., Babu, S.S.: Predicting Academic performance of students using a hybrid data mining approach. J. Med. Syst. **43**, 1–15 (2019)
9. Waheed, H., Hassan, S.U., Aljohani, N.R., Hardman, J., Alelyani, S., Nawaz, R.: Predicting academic performance of students from VLE big data using deep learning models. Comput. Hum. Behav. **104**, 106189 (2020)
10. Sekeroglu, B., Dimililer, K., Tuncal, K.: Student performance prediction and classification using machine learning algorithms. In: Proceedings of The 2019 8th International Conference on Educational and Information Technology, pp. 7–11, Mar. 2019
11. Kemper, L., Vorhoff, G., Wigger, B.U.: Predicting student dropout: a machine learning approach. European J. Higher Edu. **10**(1), 28–47 (2020)
12. Mbunge, E., Batani, J., Mafumbate, R., Gurajena, C., Fashoto, S., Rugube, T., ... Metfula, A.: Predicting student dropout in massive open online courses using deep learning models-a systematic review. In: Computer Science On-Line Conference, pp. 212–231. Springer International Publishing, Cham (2022)
13. Mnyawami, Y.N., Maziku, H.H., Mushi, J.C.: Enhanced model for predicting student dropouts in developing countries using automated machine learning approach: a case of Tanzanian's Secondary Schools. Appl. Artif. Intell. **36**(1), 2071406 (2022)
14. Bargmann, C., Thiele, L., Kauffeld, S.: Motivation Matters: Predicting Students' Career Decidedness and Intention to Drop out After the First Year in Higher Education. Higher Education, pp. 1–17 (2022)
15. Realinho, V., Machado, J., Baptista, L., Martins, M.V.: Predicting Student Dropout And Academic Success. Data **7**(11), 146 (2022)
16. Hayaty, M., Muthmainah, S., Ghufran, S.M.: Random and synthetic over-sampling approach to resolve data imbalance in classification. Int. J. Artif. Intell. Res. **4**(2), 86–94 (2020)
17. Fernández, A., Garcia, S., Herrera, F., Chawla, N.V.: SMOTE for learning from imbalanced data: progress and challenges, marking the 15-year anniversary. J. Artif. Intell. Res. **61**, 863–905 (2018)
18. He, H., Bai, Y., Garcia, E.A., Li, S.: ADASYN: adaptive synthetic sampling approach for imbalanced learning. In: 2008 IEEE International Joint Conference on Neural Networks (IEEE World Congress On Computational Intelligence), pp. 1322–1328. IEEE, June 2008
19. Rigatti, S.J.: Random forest. J. Insur. Med. **47**(1), 31–39 (2017)
20. Suthaharan, S., Suthaharan, S.: Support vector machine. Machine Learning Models and Algorithms for Big Data Classification: Thinking With Examples for Effective Learning, pp. 207–235 (2016)
21. Zhang, S., Li, X., Zong, M., Zhu, X., Cheng, D.: Learning K For KNN classification. ACM Trans. Intell. Syst. Technol. (TIST) **8**(3), 1–19 (2017)
22. Berrar, D.: Bayes' theorem and Naive Bayes classifier. Encycl. Bioinf. Comput. Biol.: ABC Bioinf. **403**, 412 (2018)
23. Suthaharan, S., Suthaharan, S.: Decision tree learning. Machine Learning Models and Algorithms for Big Data Classification: Thinking With Examples for Effective Learning, pp. 237–269 (2016)

24. Wang, W., Sun, D.: The improved Adaboost algorithms for imbalanced data classification. Inf. Sci. **563**, 358–374 (2021)
25. Bentéjac, C., Csörg, A., Martínez-Muñoz, G.: A comparative analysis of gradient boosting algorithms. Artif. Intell. Rev. **54**, 1937–1967 (2021)
26. Janiesch, C., Zschech, P., Heinrich, K.: Machine learning and deep learning. Electron. Mark. **31**(3), 685–695.eps (2021)

Data Processing and Feature Engineering for Stock Price Trend Prediction

Nguyen Thi Huyen Chau and Trung Phong Doan

Abstract In recent years, there has been a significant advancement in AI technologies and their widespread application across various aspects of life. Analyzing stock prices using machine learning techniques and data mining has become a research topic that has captured the attention of the scientific community. Stock price prediction could greatly facilitate the investors' appropriate decisions, improves profitability and hence decreases possible losses. This study aims at forecasting stock price movements using Vietnamese stock price data. We collected real-world price data of 18 stock codes from the Hanoi Stock Exchange platform. After checking for potential issues and having discovered critical data inaccuracies, we then applied various techniques to clean and augment the curated dataset. Employing technical analysis theory based on technical indicators, we proposed and implemented new features that improved prediction results by 9–20% over conventional feature generation methods. Our results are also of 2–11% better than those of some related studies on the same topic. Lastly, our data processing method achieved an accuracy of 75–80% in predicting results for the subsequent 6 months.

Keywords Stock market · Data science · Technical indicator · Data processing · Data cleaning

1 Introduction

Research employing artificial intelligence techniques to predict stock prices is increasingly becoming an appealing challenge that captures the scientific community's attention. However, in many studies, the examination and evaluation of data-

N. T. Huyen Chau (✉) · T. P. Doan
Artificial Intelligence Laboratory, Thang Long University, Nghiem Xuan Yem Road, Hoang Mai District, Hanoi, Vietnam
e-mail: huyenchau@thanglong.edu.vn

T. P. Doan
e-mail: phongdt@thanglong.edu.vn

N. Hoang Phuong et al. (eds.), *Machine Learning and Other Soft Computing Techniques: Biomedical and Related Applications*, Studies in Systems, Decision and Control 543, https://doi.org/10.1007/978-3-031-63929-6_17

related issues have yet to be addressed. One significant challenge is that stock price data typically belong to very large datasets, demanding systematic and comprehensive validation methods. Data errors, can, on the other hand, have a significant impact on the bias of the prediction results.

Investors often make investment decisions based on various information sources such as news, price history, and the company's financial statements. Stock analysis techniques can be classified into two types: fundamental analysis and technical analysis. Fundamental analysis measures a security's intrinsic value by examining related economic and financial factors. Technical analysis, on the other hand, is a trading discipline used to evaluate investments and identify trading opportunities by analyzing statistical trends gathered from trading activity, such as price movement and volume. It is very difficult to predict a specific price within a day, and doing so does not provide much useful information. Given this premise, research interest has been shifted on handling data problems to forecast future price trends.

In recent years, various techniques of Artificial Intelligence (AI) and data science have been tremendously applied in finance. Research applying machine learning combined with other techniques such as statistical analysis and natural language processing has brought in numerous achievements [1]. Stock forecasting using text mining and fundamental analysis has become an attractive field of research. Some studies have solved the stock prediction problem by recommended stock analysis based on various types of data, such as blog articles and comments [2], tweets [3], short news [4, 5], price and financial news [6].

Technical analysis is a technique widely used in AI stock forecasting [7–9]. Good results have been shown in such works as Exponential Moving Average (EMA), Weighted Moving Average (WMA), Simple Moving Average (SMA), Relative Strength Index (RSI), Money Flow Index (MFI), Moving average convergence divergence (MACD) [10–13]. The research by Zheng et al. used 17 technical indicators, 3 machine learning algorithms and a single neural network to predict the price of the next **n** days based on the previous **m** days [14]. Not only technical indicators but also oil rates, gold & silver rates, interest rate, foreign exchange rate and news could be used to good effect in [15]. Some articles and thesis studied Vietnamese stocks research to predict price trends using technical indicators [6, 16, 17]. In their research, Khoa and co-authors [18] employed technical indicators spanning 365 days to forecast the subsequent day's trends, integrating this approach with the Support Vector Machine (SVM) model to attain a noteworthy training accuracy of 96.6%. In these studies on Vietnamese stock data, there has been no mention of data corruption or data cleaning.

In this paper, we proposed a solution for cleaning stock price data and to design new features based on technical indicators. We have built AI models to evaluate the effectiveness of this method of new features construction. We also used the price of the previous n days to predict the price trend of the next m days with a sliding window of 5 days. Rather than using general indicators that apply to the entire market, our approach utilizes data specifically derived from stock codes within a particular industry sector.

Our paper are organized as follows: Sect. 2 presents our data curation. We checked potential issues and suggest cleaning data methods in Sect. 3. Section 4 describes how we generate new features from raw data, of which our hyper-parameter model is then elaborated as in Sect. 5. The experimental results are presented in Sect. 6. Last but not least, we provide a brief conclusion and perspective in Sect. 7.

2 Data Collection

We are interested in stock price data of companies in the pharmaceutical, chemical, and fertilizer manufacturing sectors listed on the Hanoi Stock Exchange. In total, the dataset pertains to 18 stock symbols: CPC, DHT, DP3, DVG, HDA, HVT, LAS, LDP, MED, MKV, NET, NFC, PCE, PGN, PMB, PMC, PPP, SDN. We collected data from the date these companies were listed on the stock exchange until July 30, 2021.

Data is collected from two websites cophieu68.com and vndirect.com.vn using frameworks in Python. For the cophieu68.com site, we used BeautifulSoup framework and scrapped data from the HTML code of cophieu68.com page. Meanwhile, vndirect.com.vn is scrapped by sending direct requests to the site's Application Programming Interface (API).

The data collected from the cophieu68 page consists of 14 attributes: *Ticker*, *DTYYYYMMDD*, *OpenFixed*, *HighFixed*, *LowFixed*, *CloseFixed*, *Volume*, *Open*, *High*, *Low*, *Close*, *VolumeDeal*, *VolumeFB*, *VolumeFS*. Meanwhile, data from the vndirect page consists of 7 attributes: *date*, *open*, *high*, *low*, *close*, *nmVolume*, *nmValue*. In this research, we are interested in only 6 attributes, namely: Trading date, opening price, closing price, highest price, lowest price and trading volume. The difference in numbers of data samples curated from the two sites is shown in Fig. 1.

3 Data Preparation

In this section, we describe the different types of problems found in the scrapped data and also present our methods to handle these issues. Checking and cleaning up data issues not only helps our data become more reliable but also facilitates the next processing step. One of the significant challenge in this phase is that we still need to ensure data continuity instead of simply deleting problematic samples or columns. In order to do so, besides discovering data inaccuracies, we also need to design effective methods of data augmentation and data substitution.

First, we can see from the Fig. 1 that the dataset from cophieu68 is smaller than that of vndirect. Through closer examining, we found out that it is because cophieu68 only stores transaction dates (i.e., days whose transaction volume is greater than 0). In contrast, vndirect's dataset stores data of all days, even those with no transactions. To

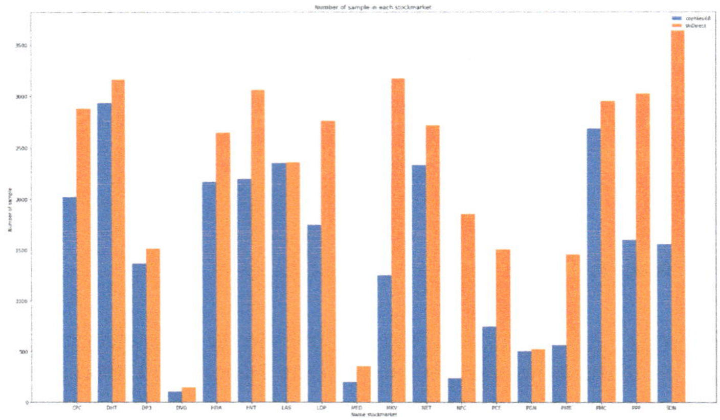

Fig. 1 Difference in the number of data samples between Cophie68 and Vndirect

Table 1 Number of rows with missing values

Stock symbols	Volume	Open	Close	High	Low
CPC	0	2	0	0	0
DP3	0	1	0	0	0
HDA	0	1	0	0	0
HVT	0	3	0	0	0
LDP	0	2	0	0	0
NET	0	7	0	1	1
PMC	0	1	0	0	0
PPP	0	3	0	0	0
SDN	0	9	0	0	0

ensure data continuity, we hence selected 12 stocks whose numbers of data samples are greater than 1000 and also choose the dataset of cophieu68 which consists of only trading days' information.

There also are 9 stocks with missing data problems in the fields of opening price, high price and low price, which all are shown in the Table 1.

– For the opening prices of the day, since their missing values are usually equal to the previous day's closing prices, we replace the missing values in this column with the previous day's closing prices. Our implementation are as follows:

$$P_{open}[i] = P_{close}[i-1]$$

– For the highest prices of the day, the missing values represent the highest prices of the day when the transaction occurred. We substitute them with the largest values of the closing prices or the opening prices. We apply the same method for missing

Table 2 Stock codes with logic errors

Stock tickets	Lowest price error	Highest price error
CPC	1	1
DHT	1	0
DP3	1	1
HVT	1	0
LDP	1	0
NET	3	1
PPP	1	0
SDN	1	5

values in the lowest prices of the day. Our implementation are as follows:

$$P_{high}[i] = max(P_{close}[i], P_{open}[i])$$

$$P_{low}[i] = min(P_{close}[i], P_{open}[i])$$

Finally, after close examination, we also discovered a logic error in the dataset. There are samples in which either the closing price or the opening price is greater than the highest price of the day or lower than the lowest price of the day. Table 2 summarize a list of stock codes subjected to this type of error.

For this type of error, we perform the process of replacing the highest prices with the largest values of either the closing or the opening prices of the day. We then apply the same process to the lowest prices of the day. Our implementation are as follows:

$$P_{high}[i] = max(P_{close}[i], P_{open}[i])$$

$$P_{low}[i] = min(P_{close}[i], P_{open}[i])$$

We performed some additional checks and found no additional errors in the dataset.

4 Feature Engineering

In this section, we design and implement new features based on the technical indicators with a time interval of 5 days between two consecutive data samples.

The construction of our new features is as follow:

Momentum shows the rate of change in price movement over a period of time. We measure the change in the closing price of the current day compared to the closing price of n days ago.

$$Momentum = \frac{C_n}{C} \times 100 \tag{1}$$

where C_n is the closing price of $\boldsymbol{n}$ days ago whereas C is the closing price of the current day.

Stochastic Oscillator compares a stock's closing price within a specific time frame. Its value fluctuates between 0 and 100. If the value goes above 80, it indicates an overbought condition, suggesting that the price may reverse. Conversely, if the Stochastic Oscillator falls below 20, it indicates an oversold condition, suggesting that the price may reverse.

$$\%K = \frac{C - L_n}{H_n - L_n} \times 100 \tag{2}$$

where $\%K$ is the Stochastic Oscillator indicator, L_n is the lowest price of n day ago; H_n is the highest price of the current day.

MACD (Moving Average Convergence Divergence) index is a trend-following momentum indicator which reflects the relationship between the Exponential Moving Averages (EMA) and the price line. Its values typically oscillate around the zero axis. When the MACD is significantly above the zero axis and begins to weaken while the price is increasing, it signals a potential peak in the stock price. Conversely, when the MACD is significantly below the zero axis and starts to rise while the price is decreasing, it signals a potential trough in the stock price.

$$EMA(n) = (C_n \times 2/(1+n)) + (EMA(n-1) \times (1 - 2/(1+n))) \tag{3}$$

$$MACD = EMA(5) - EMA(10) \tag{4}$$

Algorithm 1 Generate MACD feature

```
function EMA(C, i)
  if i == 1 then return C[i];
  return (C[i] × 2/(1 + i)) + (EMA(C, i − 1) × (1 − 2/(1 + i)))
function MACD(C,short,long, step,l)
  MACD_id = [ ]
  for i = long; i ≤ l − n2; i += step do
    C_short = [ C[i − short + 1], ..., C[i − 1],C[i] ]
    C_long = [ C[i − long + 1], ..., C[i − 1],C[i] ]
    MACD_id ← (EMA(C_short,short), EMA(C_long,long))
  return MACD_id
```

RSI (Relative Strength Index) is an indicator used to measure recent price changes, aiming to assess overbought or oversold conditions. Its values range from 0 to 100. When this index is greater than 70, it indicates that the stock is overvalued and may lead to a price reversal trend. Conversely, when the index is less than 30, it suggests that the stock is undervalued and may lead to a price reversal trend.

$$SMMA(\mathrm{S,n}) = S_n + (n-1) \times SMMA(\mathrm{S,n\text{-}1})/n \tag{5}$$

$$RSI = 100 - 100/(1 + SMMA(\mathrm{U,n})/SMMA(\mathrm{D,n})) \tag{6}$$

where U represents the set of closing prices that have increased over the past $\boldsymbol{n}$ days; D represents the set of closing prices that have decreased over the past $\boldsymbol{n}$ days.

Algorithm 2 Generate RSI feature

```
function SMMA(C, i)
  if i == 1 then return C[i];
  return (C[i] + SMMA(C, i − 1) − 1)/i
function RSI(C,n,step,l)
  RSI_id = [ ]
  for i = n; i ≤ l − n; i += step do
    U = [ ], D = [ ], RS = 0
    for j = i − n + 1; j ≤ i; j += 1 do
      U ← max(0, C[i] − C[i − 1])
      D ← max(0, C[i − 1] − C[i])
    if SMMA(D,n) == 0 then RS = 100
    else
      RS = SMMA(U, n)/SMMA(D, n)
    RSI_id ← (100 × 100/(1 + RS))
  return RSI_id
```

Bollinger Bands Width (BBW) and Bollinger %B are indicators used to measure the difference between the upper and lower Bollinger Bands, aiming to gauge market volatility:

- BBW (Bollinger Bands Width) decreases when it reflects a decrease in market volatility, and conversely, it increases when it reflects an increase in market volatility. It essentially quantifies the width of the Bollinger Bands, which expand during volatile periods and contract during less volatile periods.
- Bollinger %B describes the proximity of the latest closing price to the Bollinger Bands. When the Bollinger %B line is above 1, it indicates that the price is in an overbought state. Conversely, when the Bollinger %B line is below 0, it indicates that the price is in an oversold state. It provides insight into whether the current price is closer to the upper or lower Bollinger Bands, suggesting potential overbought or oversold conditions.

$$MA = \frac{1}{n} \times \sum_{i=1}^{n} C_i \tag{7}$$

$$BB = \sum_{i=1}^{n} (C_i - MA)^2 \tag{8}$$

$$UpperBand = MA + \sqrt{\frac{BB}{n}} \tag{9}$$

$$LowerBand = MA - \sqrt{\frac{BB}{n}} \tag{10}$$

$$BBW = \frac{UpperBand - LowerBand}{MA} \tag{11}$$

$$\%B = \frac{C - LowerBand}{UpperBand - LowerBand} \tag{12}$$

Algorithm 3 Construct BBW and Bollinger %B features

```
function BOLLINGER(C,n,step,l)
  Bollinger_id = [ ]
  for i = n; i ≤ l − n; i += step do
    B_b = 0, BB = 0                                    ▷ B_b is Bollinger %B
    MA = sum(C[i − n + 1], ..., C[i − 1], C[i])/n
    for j = i − n + 1; j ≤ i; j += 1 do
      B_b += (C[j] − MA)^2
    UpperB = MA + √(BB/n), LowerB = MA - √(BB/n)
    if UpperB ≠ LowerB then B_b = (C[i]-LowerB)/(UpperB-LowerB)
    BBW = (UpperB-LowerB)/MA
    Bollinger_id ← (UpperB, MA, LowerB, BBW, B_b)
  return Bollinger_id
```

MFI (Money Flow Index) reflects the trading flow of a stock, providing signals about price trend changes. Its values range from 0 to 100. When the index is greater than 80, it indicates that the stock is being heavily bought. Conversely, when the index is less than 20, it indicates that the stock is being heavily sold. MFI helps identify potential overbought or oversold conditions based on the trading activity surrounding the stock.

$$MF = \frac{H + L + C}{3} \times Volume \tag{13}$$

$$PMF = \sum_{\substack{i=1 \\ \mathrm{MF}_i \geq \mathrm{MF}_{i-1}}}^{n} \mathrm{MF}_i \tag{14}$$

$$NMF = \sum_{\substack{i=1 \\ \mathrm{MF}_i \leq \mathrm{MF}_{i-1}}}^{n} \mathrm{MF}_i \tag{15}$$

$$MFI = \frac{PMF}{PMF + NMF} \times 100 \tag{16}$$

where L, H is the lowest, the highest prices of the day and *Volume* is the stock's volume of the day.

Algorithm 4 Generate MFI feature

```
function MFI(C, H, L, Volume, n, step, l)
  MFI_id = [ ]
  for i = n; i ≤ l − n; i += step do
    MF = [ ], PMF = 0, NMF = 0;
    for j = i − n; j ≤ i; j += 1 do
      if j == 0 then
        MF ← 0;
        continue;
      MF ← ((H[j] + L[j] + C[j])/3 × Volume)
    for j = i − n + 1; j ≤ i; j += 1 do
      if MF[j] ≥ MF[j − 1] then PMF += MF[j]
      else
        NMF += MF[j]
    MFI_id ← (PMF/(PMF + NMF) × 100)
  return MFI_id
```

In addition to building features about technical indicators, we also built the ratio of price change between two consecutive days and the trading volume of stocks within 10 days. The ratio of the closing price, opening price, highest price, and lowest price change between two consecutive days will be in the range of $\pm 0 - 100\%$.

We have designed and implemented a total of 65 features, comprising:

- 40 features representing the price change ratio between two consecutive days within a 10-day period for closing prices, opening prices, highest prices, and lowest prices.
- 10 features representing the trading volume over a 10-day period.
- 11 features representing various technical indicators within the last 10 days, including Momentum, MACD (consisting of 2 EMA features for EMA(5) and EMA(10)), Stochastic Oscillator, RSI, MFI, and 5 Bollinger Bands-related features.

In addition to the technical indicators within the last 10 days, we have included 4 corresponding features for the technical indicators within the last 5 days, namely Momentum, Stochastic Oscillator, RSI, and MFI. The number of data samples after constructing these features has been summarized in Table 3.

To improve efficiency, we have performed features normalization using the following methods:

- **Min-Max Scaling**: We applied this method to features with values ranging from 0 to 100. Min-Max scaling transforms these features so that their values fall within a range of 0 to 1, with 0 representing the minimum value and 1 representing the maximum value.
- **Z-Score Standardization**: For the remaining features, we used Z-score standardization. This method scales features to have a mean of 0 and a standard deviation of 1, which helps maintain consistent scales and mitigates the influence of outliers.

By applying these normalization techniques, the features are transformed to have uniform scales, which can be beneficial for various data analysis and machine learning algorithms.

Table 3 The number of labels of the ticker symbols

Symbol	Sample	Increase	Decrease
CPC	403	202	201
DHT	586	278	308
DP3	272	165	107
HDA	433	203	230
HVT	438	231	207
LAS	468	218	250
LDP	348	159	189
MKV	249	114	135
NET	465	228	237
PMC	535	269	266
PPP	319	156	163
SDN	311	154	157

$$min - max = \frac{x - min}{max - min} \tag{17}$$

$$Z - score = \frac{x - \overline{x}}{\sigma} \tag{18}$$

where x is the original value before the normalization, *min*, *max* is the minimum and maximum values, $\overline{x}$ is the average value and σ is the variance of that feature in the dataset.

5 Model Development

This section will outline the process of building machine learning models to evaluate the effectiveness of the data preprocessing methods we proposed. We have chosen two models: Random Forest and Recurrent Neural Networks (RNN). It is important to note that the experiments here aim to assess the effectiveness of the data preprocessing methods rather than the performance of the machine learning models themselves.

We split our dataset by the following ratios:

- Model using Random Forest: Stock codes with sample size greater than 300 are split into training set and testing set respectively by rates of 70% and 30%. For stock codes with sample size smaller than 300, they are split are split into the training set and the testing set respectively by rates of 65% and 35%.
- Model using Recurrent Neural Network (RNN): Stock codes with sample size greater than 400 are split into training set, validation set and testing set respectively

Table 4 Results of the two models

Symbol	Random forest			RNN		
	Accuray training	Accuracy (%)	F1_score (%)	Accuray training (%)	Accuracy (%)	F1_score (%)
CPC	95.74	70.24	70.24	74.47	63	62
DHT	90.48	71.59	71.45	71.95	69	69
DP3	99.43	76.04	73.98	77.84	**75**	75
HDA	92.73	76.15	75.91	75.55	70	70
HVT	91.83	75	74.92	72.88	71	71
LAS	92.66	75.88	75.1	77.06	67	67
LDP	98.35	71.42	70.46	67.08	62	61
MKV	99.98	**79.54**	**78.18**	72.05	66	65
NET	95.38	75	74.62	67.08	63	63
PMC	92.78	72.04	72.04	71.93	71	71
PPP	99.55	70.83	70.78	65.97	65	65
SDN	97.23	75.53	75.3	75.74	72	72

by rates of 70%, 10% and 20%. For stock codes with sample size smaller than 400, they are split into training set, validation set and testing set respectively by rates of 60%, 10% and 30%.

Regarding the network architecture, the Random Forest model is constructed using the Bagging method with the following parameters: the maximum depth of each tree is 20; the maximum number of leaves in each tree is 21; the number of samples used to build a tree is 100; the information criterion is Entropy; and the number of trees is 1000.

Next, the RNN model is constructed with the following architecture: the input layer has 65 features; the optimizer is SGD (Stochastic Gradient Descent) [19]; there are two hidden layers of 32 nodes and 16 nodes, respectively.

6 Experimental Results

The results after training and building the models are summarized in Table 4. The Random Forest model achieved the best results with an accuracy of 79.54% on the testing set for the stock code MKV. For other stock codes, the Random Forest model also achieved an accuracy of over 70% on the testing set. The RNN model performed best on the stock code DP3 with an accuracy of 75% on the testing set. For other stock codes, this model also achieved accuracy of over 60% on the testing set.

Table 5 Model results with raw data

Symbol	Random forest			RNN		
	Accuray training (%)	Accuracy (%)	F1_score (%)	Accuray training	Accuracy (%)	F1_score (%)
CPC	97.16	61.98	61.98	95.04	52	52
DHT	92.92	59.65	59.07	74.63	50	50
DP3	99.6	63.54	58.65	99.43	54	55
HDA	93.39	64.61	64.07	67	46	46
HVT	93.46	62.12	60.82	79.74	55	55
LAS	92.66	68.79	68.21	81.65	54	54
LDP	96.70	63.80	61.63	65.87	57	51
MKV	100	69.31	66.62	69.54	56	54
NET	92.92	52.85	52.81	61.23	55	54
PMC	90.37	62.11	62.01	72.19	56	55
PPP	98.65	62.5	61.42	87.43	47	46
SDN	98.15	56.38	56.13	59.68	57	57

6.1 Feature Engineering and Non-feature Engineering

To evaluate the quality of our data processing method, we compared the results of the model with that of the original data before the processing. The results are summarized in Table 5.

It is evident that the results from the data processed using our proposed methods are better than the results from raw data by a margin ranging from 9 to 15%. For example, the accuracy of the Random Forest model in Table 4 ranges from 70.24 to 79.54%. In contrast, the results with unprocessed data in Table 5 are significantly lower, ranging from 52.85% to 69.31%. Similarly, for the RNN model, unprocessed data yields results ranging from 46 to 57%, which is considerably lower than the accuracy achieved with processed data, ranging from 62 to 75%.

6.2 Predicting Future Data

Our data was collected up to July 30, 2021. To assess the results of our methods, we conducted stock price trend predictions for the months of August, September, October, November, December 2021 and January 2022 for the stock codes CPC, DHT, LDP, PPP, achieving high accuracy in all cases, as can be seen in Table 6

Table 6 Summary of the forecasting results in the next 6 months

Stock symbol	Predict results (%)
CPC	76.92
DHT	76.94
LDP	84.61
PPP	77

6.3 Comparison with Results from Related Works

Our data processing method applies technical indicators in the analysis of price trends from the past 10 days to predict price trends for the next 5 days. This research bears similarities to the study of Zheng [14], which achieved accuracy of range from 41.01% to 69.5% accuracy. The commonality between both studies is the utilization of technical indicators in stock analysis to predict price trends. In contrast, the research referenced in [20] yielded a modest success rate of 52.75% using a combination of SVM, Random Forest, and Artificial Neural Networks (ANN) models with technical indicators applied to the Chinese A-share market. Notably, contemporary studies largely concentrate on forecasting aggregate market indices rather than delving into the nuances of individual stock codes, posing considerable obstacles in effectively evaluating and comparing outcomes.

Our methodology, specifically designed to align with the unique characteristics of the Vietnamese stock market, has proven effective for distinct stock codes. This is highlighted in contrast to the findings of author Khoa [18], whose study reached a peak accuracy of 96% in the training phase, whereas our approach consistently surpasses this, averaging over 96% in training outcomes. Nevertheless, this research omits any mention of the outcomes on the test dataset. We have obtained better machine learning results with accuracy ranging from 70.24% to 79.54% for the Random Forest model and from 62% to 75% for the RNN model. This demonstrates that our data processing method is suitable and effective for analyzing data in the Vietnamese stock market (Table 7).

7 Summary

In this study, we have identified issues with the current stock price dataset and proposed solutions to address them. The analysis results on the processed data have shown significant improvements compared to those on raw data. In addition to data processing, this work also suggests the design of new features based on technical analysis theory and technical indicators. This feature construction has led to good results on both the Random Forest and RNN models. Also, we have conducted comparative evaluations with related studies and demonstrated improvements of

Table 7 Summary of testing results from related research

Research article	Results	Method
Zheng [14]	69.5%	Using 17 technical indicators and Support Vector Machine
Usmani et al. [15]	77%	The model includes Oil rates, Gold & Silver rates, Interest rate, Foreign Exchange rate, NEWS
Yuan et al. [20]	52.75%	Using technical indicators, fundamental variables for Chinese A-share market
Dung and Kiem [6]	68.75%	Using news and price history
This study	**79.54%**	Using technical indicators from 10 days ago

2 to 11% of accuracy. Nevertheless, applying our method to predict price trend for the next 6 months on some stocks has shown promising results. A distinctive feature of our research is its successful implementation with individual stock codes in the Vietnamese market, diverging from the common approach of many studies that concentrate on broad market indices. In the future, we plan to explore other data processing techniques, integrate advanced analytics techniques in stock analysis to construct more features, and utilize more modern neural network models to enhance stock price trend prediction results. Predicting stock price trends is considered a complex problem that requires a combination of multidisciplinary knowledge and technology.

References

1. Mokhtari, S., Yen, K.K., Liu, J.: Effectiveness of artificial intelligence in stock market prediction based on machine learning. Int. J. Comput. Appl. (2021)
2. Bhat, A.A., Kamath, S.S.: Automated stock price prediction and trading framework for nifty intraday trading. In: 2013 Fourth International Conference on Computing, Communications and Networking Technologies (ICCCNT), pp. 1–6. IEEE (2013)
3. Makrehchi, M., Shah, S., Liao, W.: "Stock prediction using event-based sentiment analysis. In: IEEE/WIC/ACM International Joint Conferences on Web Intelligence (WI) and Intelligent Agent Technologies (IAT), vol. 1, pp. 337–342. IEEE (2013)
4. Tran, T., et al.: Neu-stock: stock market prediction based on financial news. In: Proceedings of the 2nd International Conference on Human-centered Artificial Intelligence (Computing4Human 2021). CEUR Workshop Proceedings, Da Nang, Vietnam (2021)
5. Duong, D., Nguyen, T., Dang, M.: Stock market prediction using financial news articles on Ho Chi Minh Stock exchange. In: Proceedings of the 10th International Conference on Ubiquitous Information Management and Communication, pp. 1–6 (2016)
6. Dung, P.X., Kiem, H.X.: Vietnamese stock market prediction using text mining. In: The 8th National Conference on Fundamental and Applied IT Research (FAIR) (2015)

7. Guang, L., Xiaojie, W., Ruifan, L.: Multi-scale RCNN model for financial time-series classification. In: CoRR (2019)
8. Liu, M., et al.: Stock Market Trend Analysis Using Hidden Markov Model and Long Short Term Memory (2021). arXiv:2104.09700
9. Agrawal, S., Manoj, J., Pillai, G.: Momentum analysis based stock market prediction using adaptive neuro-fuzzy inference system (ANFIS). In: Lecture Notes in Engineering and Computer Science, Mar. 2010
10. Oriani, F.B., Coelho, G.P.: Evaluating the impact of technical indicators on stock forecasting. In: 2016 IEEE Symposium Series on Computational Intelligence (SSCI) (2016)
11. Agrawal, M., Khan, A.U., Shukla, P.: Stock price prediction using technical indicators: a predictive model using optimal deep learning. Int. J. Recent Technol. Eng. (IJRTE) (2019)
12. Khairi, T., Mohammed, R., Ali, W.: Stock price prediction using technical, fundamental and news based approach. In: 2019 2nd Scientific Conference of Computer Sciences (SCCS), Mar. 2019
13. Khan, Z., Alin, T., Hussain, M.A.: Price prediction of share market using artificial neural network. Int. J. Comput. Appl. (2011)
14. Zheng, A.: Using AI to make predictions on stock market. In: cs229.stanford.edu
15. Usmani, M., et al.: Stock market prediction using machine learning techniques. In: The 3rd International Conference on Computer and Information Sciences (ICCOINS), Aug. 2016
16. Thanh, H., Meesad, P.: Stock market trend prediction based on text mining of corporate web and time series data. J. Adv. Comput. Intell. Intell. Inf. (2014)
17. Pham, K., et al.: Liquidity prediction on Vietnamese stock market using deep learning. Procedia Comput. Sci. (2020)
18. Khoa, B.T., Huynh, T.T.: Forecasting stock price movement direction by machine learning algorithm. Int. J. Electr. Comput. Eng. **12.6**, 6625 (2022)
19. Du, J.: The frontier of SGD and its variants in machine learning. J. Phys.: Conf. Ser. **1229**, 012046 (2019). https://doi.org/10.1088/1742-6596/1229/1/012046
20. Yuan, X., et al.: Integrated long-term stock selection models based on feature selection and machine learning algorithms for China stock market. In: IEEE Access, vol. 8, pp. 22672–22685 (2020). https://doi.org/10.1109/ACCESS.2020.2969293

Distributed Computing in Training Machine Learning Models

Nguyen Thi Huyen Chau, Hong Truong Ho, and Duc Tan Trinh

Abstract Artificial Intelligence (AI) has been becoming one of the fastest-growing technology fields today, with wide-ranging applications in various domains. However, the demand for high-performance hardware resources is becoming a barrier to entry for the AI development market. Also, minimizing model training time to deploy systems sooner is a challenge when dealing with large datasets. To address these issues, distributed computing methods have been proposed to perform model training, of which the two most common approaches are data parallelism and model parallelism. In this study, we proposed a novel data parallelism method for training machine learning models, accompanied by the design of a novel communication process between worker machines using Python sockets. Experimentations have shown that our distributed computing method significantly improves results compared to those of centralized computing, with up to 50% saving of model training time in optimal scenarios.

Keywords Machine learning · Distributed training · Parallel computing · Training optimization

1 Introduction

In recent years, we have witnessed the rise of artificial intelligence with applications in various aspects of life. However, for individuals or organizations to benefit from the breakthrough applications of this technology, they need access to sufficiently powerful computer systems capable of processing large volumes of data and handling complex models. Not everyone, however, possesses the resources to obtain these

N. T. Huyen Chau (✉) · H. T. Ho · D. T. Trinh
Artificial Intelligence Laboratory, Thang Long University, Nghiem Xuan Yem Road, Hanoi, Hoang Mai, Vietnam
e-mail: huyenchau@thanglong.edu.vn

H. T. Ho
e-mail: truonghh@thanglong.edu.vn

N. Hoang Phuong et al. (eds.), *Machine Learning and Other Soft Computing Techniques: Biomedical and Related Applications*, Studies in Systems, Decision and Control 543, https://doi.org/10.1007/978-3-031-63929-6_18

capabilities. This seems to be a barrier to entry into the growing artificial intelligence application market.

Also, thanks to the advent of the Internet and social media networks, online data is rapidly increasing in both volume and variety, driving the swift development of increasingly sophisticated AI models [1–3]. The demand for training machine learning models is on the rise, with models becoming more complex and requiring more data, leading to high requirements for algorithms and powerful computing systems. In reality, training models today often consume a significant amount of time for parameter tuning, sometimes lasting weeks [4]. To accelerate training time, systems such as OpenAI's 10,000-GPU ChatGPT training system have been developed. However, not many companies or organizations can have access to the resources of such large systems.

To address this issue, distributed machine learning systems have been researched and developed to enhance scalability at a low cost. Distributed model training can be divided into several categories [5, 6]. Most distributed training methods focus on GPU-based distributed training and how to implement this methods using frameworks [6]. In recent years, major machine learning libraries have introduced their own distributed solutions, such as Torch with torch.distribute [7], or TensorFlow with tf.distribute [8].

While these solutions are available, they often focus on building distributed computations on GPUs and may not be widely applicable to different types of machines [8–10]. In the meantime, given the constraints of our workspace at the AI Laboratory at Thang Long University, we needed to deploy a specialized solution that allows distributed training not only between GPUs but also applicable between CPU machines with diverse configurations. Therefore, we needed to design a new parallelization solution that is suitable for the context of our laboratory. The solution also had to ensure no additional costs, space, or the need to build a new computer system while utilizing the existing machines. Essentially, the solution had to support distributed computing across multiple machines, be lightweight, and highly customizable.

In this paper, we propose such a solution that applies computing to multiple computers connected via an intranet to leverage existing computers at work, contributing to accelerate the training process of machine learning models.

Our solution was developed based on two key ideas from distributed training models: data parallelism and decentralized asynchronous systems. We used Python programming language and the Socket library to program the connections between processes. Leveraging the TCP protocol, we then propose a new communication process among serving machines to optimally align with the implemented model.

The paper is organized as follows: Sect. 2 introduces an overview of the various modes and architectures of distributed training. In Sect. 3, we propose a novel distributed computing model and describe its design and implementation in detail. Section 4 then demonstrate the experimentation results of the novel model. We evaluated our model on a standard dataset and found that it can achieve significant speedups over training on a single machine. We also provide insights and experiences gained during the research, installation, and deployment of our distributed training models. Section 5 then discuss some research directions perspective. We found that our model is an effective technique for training large-scale machine learning models. However, there are still some challenges to be addressed in the future. Last but not least, we provide a brief conclusion in Sect. 6.

2 Distributed Computing Overview

2.1 Data Parallelism Versus Model Parallelism

There are two popular approaches to distributed training: model parallelism and data parallelism [11–13]. Model parallelism is a distributed training method in which each part (such as layers, blocks, or even individual operations) of the model is divided into worker machines, where each worker machine performs the calculation of one or more parts of the model (Fig. 1a). This method therefore is suitable for large models [8]. Data parallelism, on the other hand, divides the data into worker machines. Each worker machine simultaneously stores a copy of the model. This method therefore is more suitable for large datasets. In this method, each worker machine performs training on the assigned dataset to calculate the gradient (Fig. 1b). The gradient is then sent to the master machine for aggregation. This method is feasible because current machine learning methods do not create models that require interdependent datasets during training [11]. Therefore, we can approximate the gradient of a mini-batch by the sum of the gradients of the data in that mini-batch. This can be expressed by the following formula:

$$\frac{\partial \mathcal{L}(w, x)}{\partial w} \approx \frac{\partial \mathcal{L}(w, x_1)}{\partial w} + \cdots + \frac{\partial \mathcal{L}(w, x_n)}{\partial w}$$

where x being the data of the mini-batch and $(x_1, \ldots, x_n)$ being non-overlapping subsets of x [5].

We have chosen the approach of data parallelism for the following reasons: Firstly, this method is the simpler and quicker to implement, since dividing the dataset is easier than dividing each part of the model into each machine. Secondly, model parallelism does not seem to be suitable for the kind of mostly small and medium models being deployed in our laboratory research. Since subsequent computing blocks still have to wait for previous computing blocks to complete before they can continue; this method hence does not improve execution time significantly for medium and small models when the model can be loaded entirely on one machine, but is more suitable for large models when the model cannot be loaded entirely on one machine.

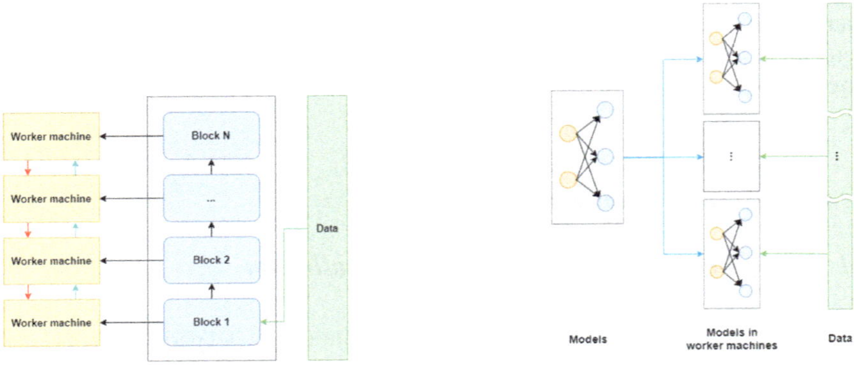

(a) Model parallelism (b) Data parallelism

Fig. 1 Model parallelism and Data parallelism

2.2 *Decentralized Asynchronous Systems*

When building a distributed system for machine learning, we often consider such factors as centralized versus decentralized and synchronous versus asynchronous modes. With centralized optimization, worker machines only perform gradient calculations and all optimization and model updates are pushed to a central master machine [5]. This method is vertically scalable, i.e., increasing the number of worker machines involved. However, there is still a drawback of vertical scaling, which is that if the number of worker machines is too large, a single center cannot handle it all and leads to a bottleneck [14, 15]. In addition, we have a decentralized system. In contrast to centralized, decentralized allows machines to have their own optimization functions and the right to update their own models. The decentralized system can scale both vertically and horizontally.

Synchronous is a method in which the model update is performed by the master after all machines have sent calculation information, weight, or gradient. The master will wait until all workers have completed their work before updating the model. When a worker machine completes the work first, it may have to wait for other machines. The master will use the calculation information to update the model. Then the master will send the model back to all machines. After receiving the new model, the machines will continue to perform the training task. It can be seen that when starting an epoch, the model on the worker machines is the same, that is, synchronous.

Asynchronous is a looser approach wherein the master machine will no longer wait for all machines. As soon as a worker machine sends calculation information, the master machine will update the model. The updated model will only be sent back to the single machine that contributed the calculation information. Only in the first epoch are the worker machines have the same model. This method helps worker machines not have to wait for each other, but the number of times the master machine updates and sends the model increases. Asynchronous has one drawback is that if a machine is running slower than other machines, it will update the central model with information from the past; this can affect the overall pull down the results of the central model.

3 Proposed Distributed Computing Method

3.1 Communication Process with Socket Library in Python

We will design our novel method of distributed computing using Data Parallelism and Decentralized Asynchronous Systems approaches. We will implement the method using Socket library of Python [16]. In order to do so, we also need to design a protocol for machines to inter-communicate, transmit data, and update the model.

3.1.1 Socket

Sockets are endpoints in a two-way communication channel. Sockets support communication within the same process, between processes on the same device, and between processes on different devices. Devices connect to each other through the hostname and port number. Sockets are typically connected in two ways: stream sockets and datagram sockets. Each type of connection has its own characteristics [17].

Datagram sockets are connectionless sockets that operate on the UDP protocol. Datagram sockets focus on sending simple, fast files, so they do not support data checking methods when sending packets. This means that data may be lost, corrupted, or not received by the receiver, even though the data transmission speed is very fast.

Stream sockets, on the other hand, are connection-oriented sockets that operate on the TCP protocol. Stream sockets focus on the accuracy of the data being transmitted. They typically establish connections in a 3-Way Handshake process and disconnect in a 4-Way Handshake process. Additionally, the TCP protocol always ensures that files are sent to the correct address and in the correct order, which helps to minimize data errors and losses.

3.1.2 Communication Process

To be suitable for training distributed models, we propose a communication process based on the TCP protocol. The process consists of 3 child processes, the Initialization process, the Data sending process, and the Model updating process.

- The initialization process (Fig. 2) is the first process that is executed and only executed once. First, the worker machine sends a connection request to the master

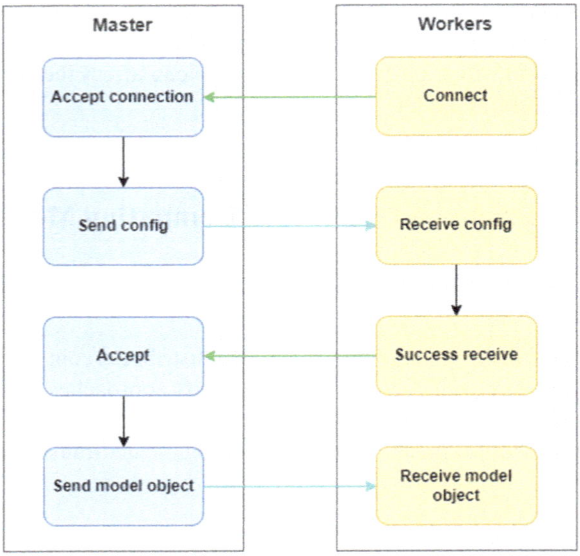

Fig. 2 Initialize model process

machine. When the master machine agrees to connect, the master machine will create a new stream to serve all subsequent activities and requests from the worker machine. After the connection has been established, the master sends the training hyper-parameter configuration to the worker machines. When the worker machine successfully receives it, the worker must send a message. This helps the master know that the worker has received the data and stops sending the data. Because TCP is a reliable protocol, TCP will send the data multiple times to ensure that the recipient will receive the packet. Next, the master sends the model with the existing parameters in the form of an object to the worker machines to ensure that the system has a common starting point. After receiving the model, there will be no step to send a message that has been received successfully as when receiving the configuration content because immediately afterwards the worker machine will have to perform the data request. Because only when the model is completed can the worker perform the data request. The first data request message after receiving the model will have the effect of letting the master know that the model has been sent successfully. Therefore, sending additional information that the model has been received is not necessary in this process.

- The data sending process (Fig. 3) is the process that distributes data segments to worker machines. Before this process begins, the master machine must build a data queue; Each element in the queue stores different parts of the data that will be distributed to the worker machines. In this process, the worker machines will send a data request to the master machine. After receiving the request, the master machine will retrieve information from the queue to send data to the worker machines. After receiving all the data, the worker machine will send a message to confirm that it

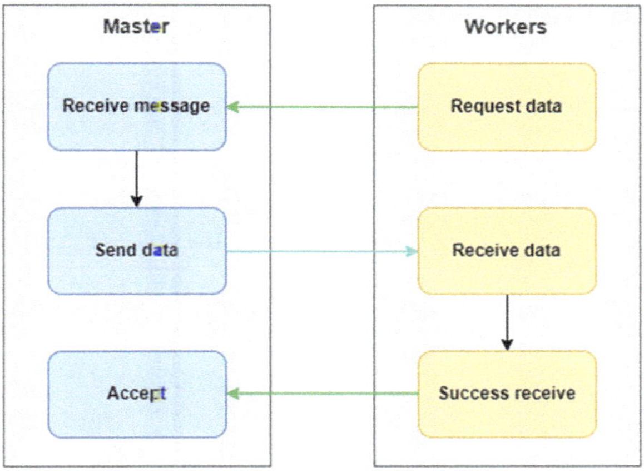

Fig. 3 Send data process

has received the data. Although we may not need to send a successful message because in the next process, the update model process, the worker will have to send a model update request message to replace the successful message, similar to the initialization process. However, we see that after completing the data receiving process, the worker machine will perform the model training process, which may take a long time. Therefore, it is necessary for the worker to send a message to confirm that it has received the data for the master machine to stop sending data and reduce the master's workload.

- The model update process (Fig. 4) is activated immediately after the workers complete the training of the mini-batch. The worker will start the sending process by sending a notification to the master. Immediately afterwards, the worker continues to send the parameters of the model that have been encoded to the master. The master decodes and updates to the original model. After completion, the master will send the new model back to the worker machines so that these machines can train data based on the latest model in subsequent training sessions. At the end of this process, the worker will send a notification to receive the model successfully. At the end of this process, the program can repeat process two, repeat process three, or terminate the program as desired. At the end of this process, there must be a confirmation step to receive the success, because if the third process is repeated, there may be problems as explained above; Or if the program terminates, we must interrupt data transmission first to avoid errors for the master or worker side.

We designed this three-process procedure so that in Sect. 3.2, depending on the data deployment requirements, we can perform the data sending process one or more times without causing other processes to fail.

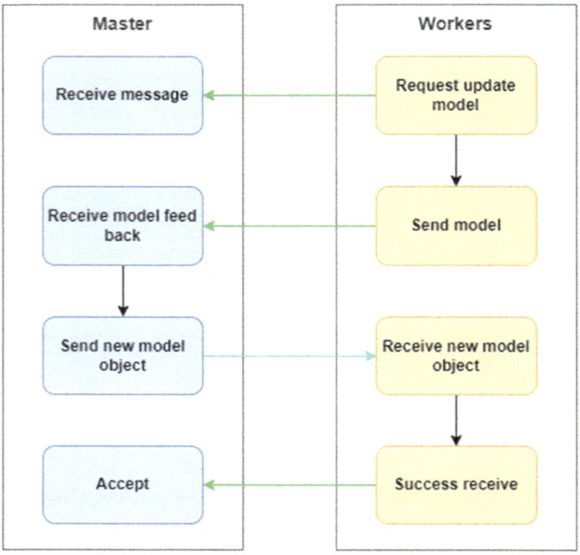

Fig. 4 Update model process

3.2 Data Parallelism Model Design and Deployment

In this section, we will introduce how we implement our data parallelism solution for decentralized asynchronous systems. With this solution, we have two options for implementing this solution, which are fixed data and continuously changing data. In the case of fixed data, the divided data is only sent once. Worker machines will automatically swap data with each training epoch. The data changes continuously with each epoch, the master machine takes data from the queue and sends the assigned data to each machine. With each method, the algorithm model is divided into two parts. The part running on the master machine and the part running on the worker machine.

Before implementing either of these methods, the master must perform the following steps. The master sets up a socket using the TCP/IP protocol with a configuration that includes the machine's IP address and port. Then, the master initializes the hyperparameter configuration and initializes the model. The hyperparameter configuration will include the following information: number of epochs, batch size, and steps per epoch. Other hyperparameters, such as learning rate and optimization method, are attached directly to the model and are sent along with the model, so they do not need to be sent in the configuration anymore. When a worker machine connects to the master, the initialization process is performed. The difference between the two methods we mentioned will be in how we implement the second and third processes.

The first method we want to mention is the fixed data sending method. This method gives us performance in terms of model training time because the data is sent once before training, so the model training process will not be affected by waiting for data sending and receiving. The training data distributed to the worker machines will be sent and stored to serve the training process behind. Finally, the data will be trained and continuously perform the process of updating the model each time the training process is completed. The detailed operating procedures of the master and worker machines during the training and updating of the model are shown in algorithms 1 and 2. It can be seen that in this method, process two is performed once and process three will be performed multiple times.

Algorithm 1 Master node handler for worker ith fixed data

Require: model weight $\hat{w}$
1: **while** still connection **do**
2: Wait weight Δw^i_j of worker i
3: $\hat{w} \leftarrow \hat{w} + \Delta w^i_j$
4: Send weight $\hat{w}$ to worker i
5: **end while**

Algorithm 2 Worker node ith training program fixed data

Require: $\mathcal{D}_i$ dataset for worker i, number of epochs k, inital weihgt w_0
1: **for** $j \leftarrow 1 \cdots k$ **do**
2: $w_{temp} = w^i_{j-1}$
3: Fit model with dataset $\mathcal{D}_i$
4: Get w^i_j from model
5: $\Delta w^i_j = w^i_j - w_{temp}$
6: Send weight Δw^i_j to master
7: Receive weight $\hat{w}$ from master
8: $w^i_j \leftarrow \hat{w}$
9: **end for**

The method of sending multiple data at once has the advantage of being fast and easy to implement. However, on the contrary, because we are implementing a decentralized asynchronous distributed computing model, there is a disadvantage: if a machine is slow or even worse, it loses connection, then a part of the data will be less or not trained, thereby reducing the overall efficiency. Therefore, we think that instead of having fixed data, we will replace the data continuously, then if a machine is slow, the machine can make up for the slowness of the other machine later by receiving the work of that machine. Therefore, the model will be read all the data.

The second method is the method of continuously changing data. The advantages of this method are to reduce the data storage capacity, ensure data information security. However, the model training time is prolonged due to interruptions by the data sending and receiving process. This method has the initialization process and the model updating process similar to the fixed data method, the difference lies in the data sending process. In the process, instead of the master sending all the data in the queue and then deleting the queue; the master will use the queue to distribute and push back the data that has been trained back into the queue. The process on the worker machines will send data request and be coordinated by the master to receive data. Each time the sub-data packages are sent successfully, worker will perform training and then perform the model update process. The process of sending and receiving data and training is repeated continuously.

Algorithm 3 Master node program for worker ith mutate data

Require: initial model weight $\hat{w}_0$, number of worker node n, number of epochs k, Dataset queue $\mathcal{Q}$

```
1: while still connection do
2:     Receive dataset D_i in worker i
3:     Q.put(D_i)
4:     D_i ← Q.pop()
5:     Send dataset D_i to worker i
6:     Wait weight Δw^i_j of worker i
7:     ŵ ← ŵ + Δw^i_j
8:     Send weight ŵ to worker i
9: end while
```

Algorithm 4 Worker node ith program mutate data

Require: number of epochs k, inital weight w_0

```
1: for j ← 1...k do
2:     w_temp = w^i_{j-1}
3:     Receive dataset D_i from master
4:     Fit model with dataset D_i
5:     Get w^i_j from model
6:     Δw^i_j = w^i_j − w_temp
7:     Send weight Δw^i_j to master
8:     Receive new weight ŵ from master
9:     w^i_j ← ŵ
10: end for
```

The advantage of the continuous data sending method is that it does not create data copies on worker machines. This reduces the amount of data storage, helps to secure data information. On the contrary, continuous data sending between machines can slow down the training process because the master machine must send data to all machines. With each epoch, the entire data will be sent once. The larger the dataset, the longer the sending time will be and therefore slow down the training process. In addition, continuous data sending will also be affected by network bandwidth. If the amount of data sent is greater than the load capacity of the network link, the time it takes to send all the data will increase.

4 Experimentations

4.1 Experimentation Results

The data used is the dogs-vs-cats dataset [18]. The dataset used for training includes 25,000 images evenly divided into 2 classes, dogs and cats. Training is performed on one master machine and 2 worker machines. In fact, a decentralized system can have multiple masters, but we want to conduct a simple experiment, so we only perform it with one master and two workers. The two worker machines will have the same configuration, including an RTX 2070 8GB GPU, and an Intel(R) Core(TM) i7-8700 CPU. We train the dataset for 10 epochs, using both the fixed data and the changing data methods.

The model we use is the ResNet-18 model [19] with the final Fully-connected layer changed to 1 neuron and sigmoid activation to fit the binary classification problem of dogs and cats. The final model will have 18 layers with 11 million parameters. Although both the model and the data are not large enough for one machine to run easily, we can still see the time saved to evaluate the effectiveness of the method.

Table 1 shows that the results of the running time of the fixed data method are overwhelming compared to the method. The time is significantly reduced, the time is reduced by half compared to running without distribution. Looking at the results, we predict that the distributed training time will be inversely proportional to the number of distributed machines. The changing data method has reduced the time but not significantly compared to the base case. However, if the machine used cannot train the entire dataset by itself, this method can still be effective.

4.2 Insights and Experiences

In this section, we provide some insights and experiences gained during the research, installation, and deployment of our distributed training models. These experiences, from successful models to failed models, will provide you with some suggestions for the model design and setup process, and when the detailed installation.

With the computers available in the organization, we choose the machine with the most CPU threads possible for the master machine. This is because the master machine will be connected to many worker machines, each of which will need a

Table 1 Comparison result

Method	Run time (10 epochs)	Average
No distribute	11 min 20 s	68 s/epoch
Fixed data	**5 min 16 s**	**32 /epoch**
Mutate data	10 min 26 s	66 s/epoch

separate processing thread to always be ready to wait for requests from the worker machines. This will ensure that the master machine is capable of managing the workers and updating the models as soon as possible. The remaining computers will take on the task of training data. Therefore, computers with powerful GPUs will be installed as worker machines. We also need to pay attention to the selection of the number of worker machines. The number of machines will depend on the single-core processing power of the CPU and the bandwidth speed of the internal network.

Sending and receiving large files is risky, so we need to split the data into smaller segments to send. When sending and receiving, the data is stored in binary format in memory. Through research and testing, we found that using buffer sizes of 1024 bytes, 1024 * 8 bytes, or 1024 * 32 bytes is effective. The specific number to choose depends on the network bandwidth where the system is set up. Typically, the buffer size should not exceed the network bandwidth, which can lead to congestion, and it should not be too low to avoid sending multiple times due to splitting the data file too small.

When receiving and merging data segments into complete data, there is also a point to note that we should use bytearray instead of bytes. Both data types allow for storing data in binary format. Although initializing data in bytes is faster because it is an immutable data type (that is, a data type that cannot change its value), the bytearray data type supports the insert and append functions to improve data merging capabilities. Therefore, the use of bytearray significantly improves performance.

Finally, it is not recommended to set the model to disconnect each time the data is sent. Although disconnecting can free up CPU threads, the time to initialize the threads and reconnect the Socket also takes a lot of time when we reconnect.

5 Discussion on Future Works

Though having good results, the current model still has such limitations as the fixed number of computers, the fixed configuration for all computers, or the limited number of experiments. We have not yet experimented with multiple models of different sizes, and we have not yet experimented with models with different numbers of worker machines. The current model is just suitable for worker machines with the same processing speed. In the near future, we plan to continue to develop our model while taking some new aspects into consideration.

We want to research, experiment, and set up on multiple configurations of computer local network to find the right number of computers for the model. This requires taking into account factors such as network speed, the processing power of the master machine, and the processing power of the worker machine to come up with the appropriate set of parameters. In addition, we are also interested in the performance achieved on the cost of hardware devices. We want to find products, test, and experiment to come up with a list of products with the best performance-to-cost ratio, from which to achieve maximum economic benefits.

Currently, the worker machines we use have the same configuration. In the future, the machines may not be that homogenic. We want to build a data partitioning method for training that is more efficient and flexible. We want to distribute data to different machines based on the performance of the computer. This will minimize the training time when machines with different performance are assigned the same amount of data, leading to machines having to wait for slower machines to finish the work. In the process of sending data, random elements can be added, such as sending random datasets, thereby increasing the objectivity when building the model.

While experimenting with another distributed model, we also realized that it is not advisable to set up each step with a dataset that is too small. This can lead to the weights of the model being sent back continuously, and we know that the process of sending data through Socket is not fast. Therefore, a bottleneck can occur, causing the training process to slow down. To choose the right batch size and number of steps, we need to calculate based on the size of the model and network speed. From the size of the model and network speed, we can calculate the time to send the model object, then test the training speed of the model and fine-tune the hyperparameters to match to get the training time to be comparable to the time to send the model.

Data security issues are also an area that needs to be focused on. With the direction of expanding the distributed model to the cloud, in the future, not only will the model be trained on computers connected in an internal network, but it can also be connected through the internet. This poses many risks of data leakage or data loss. Therefore, it is necessary to find solutions to ensure that the data is secure during the process of sending and receiving data as well as during the training process at the worker machines.

Building a hybrid model between model parallelism and data parallelism is also a development direction in parallel with the above issues. Today not only is the amount of data increasing, but machine learning models are also increasing in size. Data parallelism is mainly to solve the problem of large data, so we must also develop model parallelism models or hybrid models between the two models to be able to solve the problem in a more thorough way.

6 Conclusion

The use of distributed computing has the potential to accelerate the training process, which will then help the field of artificial intelligence develop further in the future when distributed computing can solve the barrier of machinery and equipment for organizations. In this paper, we propose a distributed training model based on data parallelism and decentralized Asynchronous System models. Our model uses sockets to connect computers on an internal network, communicating with each other according to a novel process built on top of the TCP protocol to be suitable for training machine learning models. By experimentations, our model is shown to have speed up significantly the model training process. Along with that, we summarize and provide some experiences in building distributed training models. Last but not least, we offer some perspective on our research directions. It is expected that distributed training models would be a driving force for the development of large-scale machine learning models hence for AI research and applications in the future.

References

1. Constine, J.: How big is Facebook's data? 2.5 billion pieces of con- tent and 500+ terabytes ingested every day (2012). http://techcrunch.com/2012/08/22/how-big-is-facebooks-data-2-5-billion-pieces-of-content-and-500-terabytes-ingested-every-day
2. Campbell, S.D., et al.: The explosion of artificial intelligence in antennas and propagation: how deep learning is advancing our state of the art. IEEE Antennas Propa. Mag. **63**(3), 16–27 (2021). https://doi.org/10.1109/MAP.2020.3021433
3. Erdil, E., Besiroglu, T.: Explosive growth from AI automation: a review of the arguments (2023). arXiv: 2309.11690 [econ.GN]
4. Floridi, L., Chiriatti, M.: GPT-3: its nature, scope, limits, and consequences. Minds Mach. **30**, 681–694 (2020)
5. Langer, M., et al.: Distributed training of deep learning models: a taxonomic perspective. IEEE Trans. Parallel Distrib. Syst. **31**(12), 2802–2818 (2020). https://doi.org/10.1109/TPDS.2020.3003307
6. Verbraeken, J., et al.: A survey on distributed machine learning. ACM Comput. Surv. (csur) **53**(2), 1–33 (2020)
7. Ketkar, N., Santana, E.: Deep learning with Python, vol. 1. Springer (2017)
8. Abadi, M., et al.: TensorFlow: a system for large-scale machine learning. In: 12th USENIX Symposium on Operating Systems Design and Implementation (OSDI 16), pp. 265–283 (2016)
9. Chen, T., et al.: Mxnet: a flexible and efficient machine learning library for heterogeneous distributed systems (2015). arXiv:1512.01274
10. Chilimbi, T., et al.: Project adam: building an efficient and scalable deep learning training system. In: 11th USENIX Symposium on Operating Systems Design and Implementation (OSDI 14), pp. 571–582 (2014)
11. Dean, J., et al.: Large scale distributed deep networks. In: Pereira, F. et al. (Ed.) Advances in Neural Information Processing Systems, Vol. 25. Curran Associates, Inc. (2012)
12. Krizhevsky, A., Sutskever, I., Hinton, G.E.: ImageNet classification with deep convolutional neural networks. In: Pereira, F. et al. (Ed.) Advances in Neural Information Processing Systems, vol. 25. Curran Associates, Inc. (2012)
13. Hillis, W.D., Steele, G.L.: Data parallel algorithms. Commun. ACM **29**(12), 1170–1183 (1986). ISSN: 0001-0782. https://doi.org/10.1145/7902.7903
14. Langer, M., et al.: MPCA SGD—a method for distributed training of deep learning models on spark. IEEE Trans. Parallel Distrib. Syst. **29**(11), 2540–2556 (2018)
15. Lian, X., et al.: Can decentralized algorithms outperform central- ized algorithms? A case study for decentralized parallel stochastic gradient descent. In: Advances in Neural Information Processing Systems 30 (2017)
16. Hunt, J., Hunt, J.: Sockets in Python. In: Advanced Guide to Python 3 Programming, pp. 457–470 (2019)
17. Ingalls, R.: Sockets tutorial. In: línea] (2014). http://www.cs.rpi.edu/moorthy/Courses/os98/Pgms/socket.html
18. Cukierski, W.: Dogs vs. Cats (2013). https://kaggle.com/competitions/dogs-vs-cats
19. He, K., et al.: Deep residual learning for image recognition (2015). arXiv: 1512.03385 [cs.CV]

Fruit Calorie Determination System for Dieters and Athletes Using Deep Learning

Nguyen Thai-Nghe, Ngo Quoc Vi, and Nguyen Chi-Ngon

Abstract People become busier at work in modern life, fast food is usually a choice. However, this daily habit can cause several problems such as overweight and obesity diseases. In this work, we propose a calorie determination system using deep learning approach, especially for dieters and athletes. This work firstly focuses on fruit calorie determination. First, we collect the dataset of popular fruits in Vietnam and other public sources, then we propose using deep learning with MobilenetV2 architecture for fruit classification. An application is built for automatically determining the calorie of those fruits from photo taken by smartphone. The method can achieve more than 90% accuracy and the application can work on both Android and IOS devices.

Keywords Fruit classification · Calorie determination system · Deep learning

1 Introduction

People today are busier at work, and fast food is frequently chosen. But this everyday practice can lead to a number of issues, including ailments like obesity and overweight. Fruits and vegetables are vital food in human life, it provides many nutrients, fibers and vitamins to help people stay in good physical and mental health and reduce the risk of disease. However, to be healthy, we should use fruits under control. There are so many types of fruits and vegetables in the world that make it difficult for us to remember the names and nutritional values of each. This is difficult for us to control the amount of energy we put into the body when eating certain fruit. With the great development of current technology, this problem can be solved by the emergence of fruit classification application. Build a fruits and vegetable classification system to

N. Thai-Nghe · N. Q. Vi · N. Chi-Ngon (✉)
Can Tho University, Can Tho City, Vietnam
e-mail: ncngon@ctu.edu.vn

N. Thai-Nghe
e-mail: ntnghe@ctu.edu.vn

N. Hoang Phuong et al. (eds.), *Machine Learning and Other Soft Computing Techniques: Biomedical and Related Applications*, Studies in Systems, Decision and Control 543, https://doi.org/10.1007/978-3-031-63929-6_19

determine the calorie content of each type and simultaneously help users to check the calorie content for a healthy diet.

In this work, we propose a calorie determination system using deep learning approach, especially for dieters and athletes. This work firstly focuses on fruit calorie determination, we are currently working on other kind of food such as meat and milk.

2 Related Work

Fruit recognition and classification is considered a complex task and still facing certain challenges. To develop a perfect fruit recognition and classification technique these challenges need to be overcome. This section presents the work of researchers who have attempted to solve some of these problems using various methods such as machine learning and deep learning methods.

The authors in [1] have developed a fruit classification and detection system using deep learning algorithms. They used two datasets of colored fruit photos in this study. In addition to a private dataset of 761 photographs with eight different categories of fruits that they have collected and analyzed on their own, the first FIDS-30 dataset of 971 images with 30 separate classes of fruits is publicly available. In this work, individual fruit detection across many classes has been accomplished using the YOLOv3 deep learning object detection algorithm, and the final classification for the recognition of a single category of fruit in photos has been accomplished using ResNet50 and VGG16 techniques. They then used Flask for the web framework to create the models for automatic fruit classification. With ResNet50 and VGG16, they obtained accuracy from the public dataset of 86 and 85%, respectively.

The authors of [2] present a method for developing a system for classifying fruits and vegetables in the retail market using images obtained using a video camera attached to the system. The system's objectives include reducing the quantity of human–computer interactions, accelerating the identification process, and improving the usability of the graphical user interface in comparison to current manual methods. The Raspberry Pi, camera, display, load cell, and casing make up the system's hardware. Various convolutional neural networks have been evaluated and retrained to classify an item. The authors in [3] using deep learning models such as CNN, RNN, LSTM for classification of fruits based on chosen optimal and derived features.

There are several works on determination of vitamins in fruits using different approaches. The work in [4] presented a review on fruit recognition and feature evaluation using CNN. A web-based tool for calculating fruit calories was proposed in the study cited in [5]. Using typical image datasets and numerous improvements including pre-processing, segmentation, feature extraction, training, and classification using shape and size with the use of machine learning algorithms, the calorie fraction of a fruit is identified. Estimating the number of calories in the fruit and giving it to the users is the last stage.

In this work, we propose a calorie determination system using deep learning approach, especially for dieters and athletes. This work firstly focuses on fruit calorie determination. First, we collect the dataset of popular fruits in Vietnam and other public sources, then we propose using deep learning with MobileNetV2 [6] architecture for fruit classification. An application is built for automatically determining the calorie of those fruits from photo taken by smartphone. However, other classification approaches [7, 8] could also be used.

3 Proposed Approach

The system architecture is presented in Fig. 1. This system consists of two sides: Clients are mobile devices, including Android and iOS. The server side processes images, identifies, determines the calorie content and retrieves fruit information. The client sends the image to the server, then the server will save and process it. The pillow will convert the image to vector. Then the system model will receive this vector and classify. After identifying the fruit, the system will use web-scrap to get calorie and fruit information, then return the result to the client as JSON, as presented in Fig. 2.

For classification, this work uses the MobileNetV2 [6] which is a convolutional neural network architecture that seeks to perform well on mobile devices. It is based on an inverted residual structure where the residual connections are between the bottleneck layers. Its architecture is presented in Fig. 3.

In the proposed system, the user can upload the images of fruits from album or camera. After classification, the system can determine which kind of fruits. Based on the nutrition fact as an example in Fig. 4, it automatically calculates the nutrition.

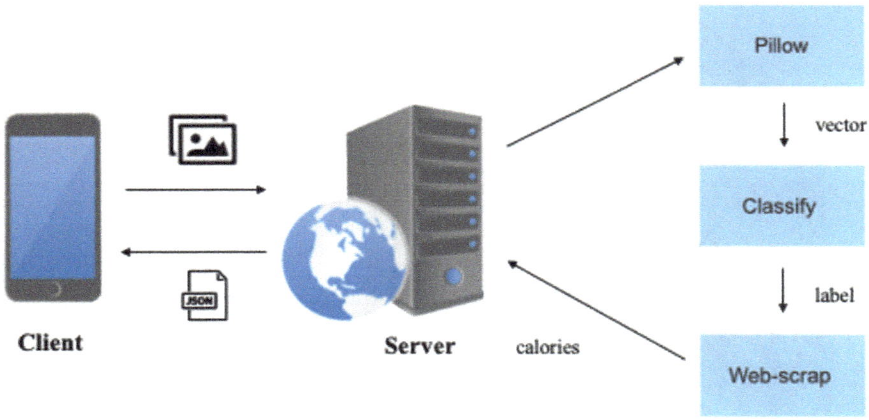

Fig. 1 System architecture

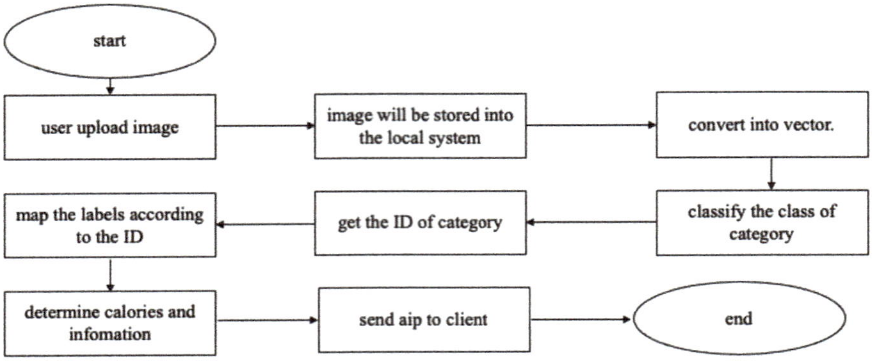

Fig. 2 Workflow diagram

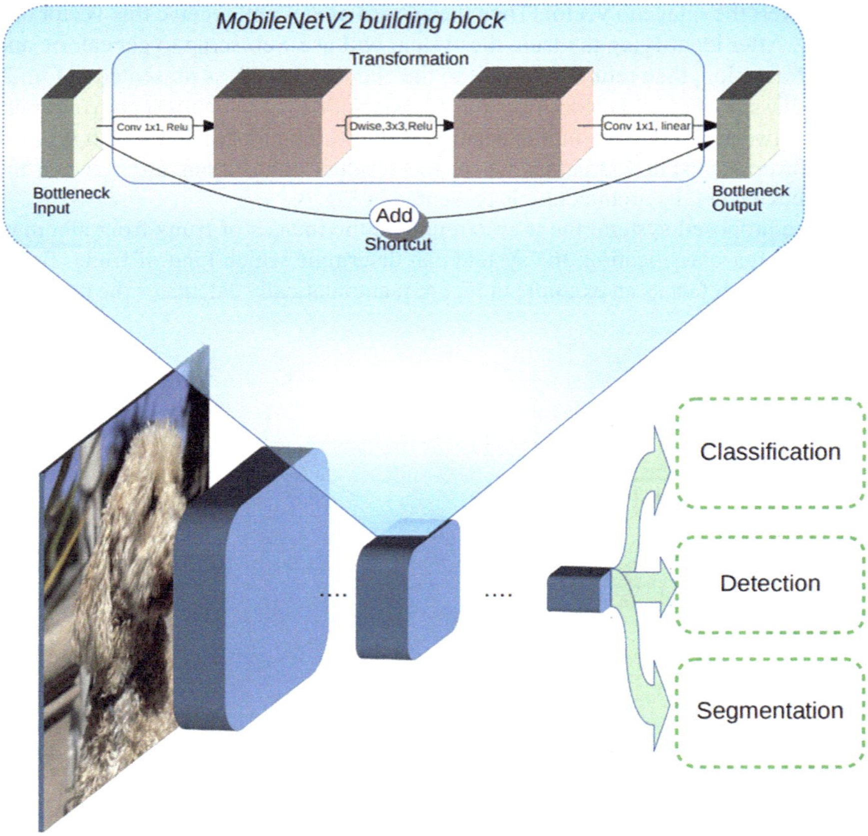

Fig. 3 MobilenetV2 architecture. *Source* https://ai.googleblog.com/2018/04/mobilenetv2-next-generation-of-on.html

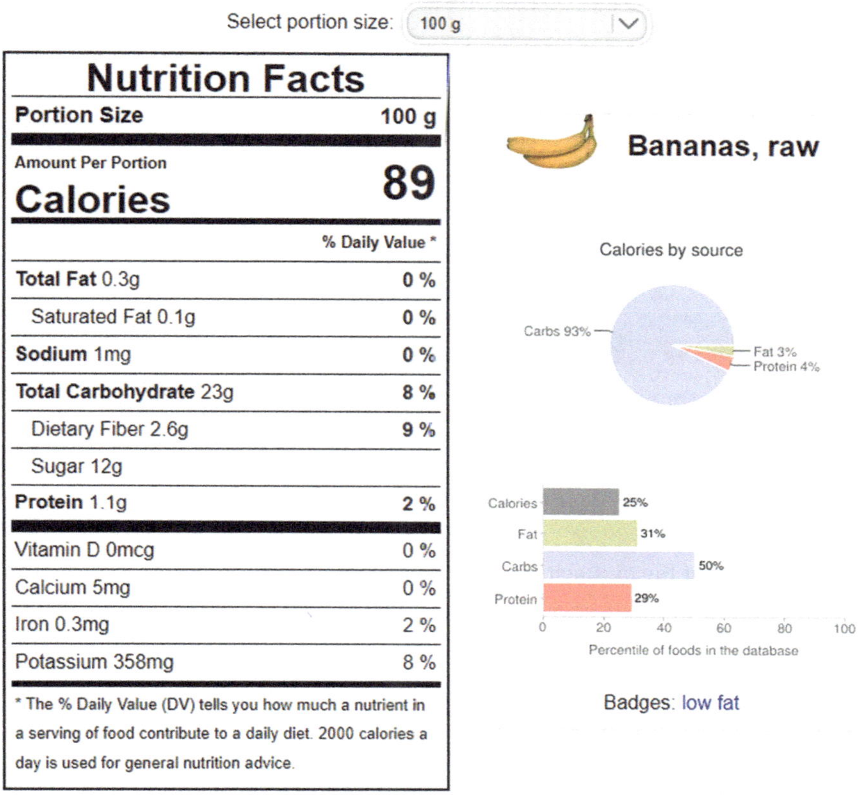

Fig. 4 Banana nutrition fact. *Source* https://www.nutritionvalue.org

4 Experiments

4.1 Dataset

We have used the "Fruit and Vegetable Image Recognition" dataset[1] for experiments. This dataset has 36 classes of fruits (banana, apple, pear, grapes, orange, kiwi, watermelon, pomegranate, pineapple, mango) and Vegetables (cucumber, carrot, capsicum, onion, potato, lemon, tomato, raddish, beetroot, cabbage, lettuce, spinach, soy bean, cauliflower, bell pepper, chilli pepper, turnip, corn, sweetcorn, sweet potato, paprika, jalepeno, ginger, garlic, peas, eggplant) as presented in the left side of Fig. 5.

Moreover, we have added 6 popular fruits in Vietnam to the dataset including of coconut, custard apple, guava, persimmon, pumpkin and winter melon, as presented

[1] https://www.kaggle.com/datasets/kritikseth/fruit-and-vegetable-image-recognition.

Fig. 5 Dataset for experiments

in the right side of Fig. 5. Each type of fruit has 100 images for train, 10 images for validation, and 10 images for test.

4.2 Experimental Results

Figure 6 presents the accurary and loss during training and validation process. Results show that the classification model can work well. The final ccuracy on the test set achieve 93.48% of accuracy.

After training the model, we build an application for fruit calorie determination as presented in Fig. 7. In this system, the users select a photo of fruit from the camera

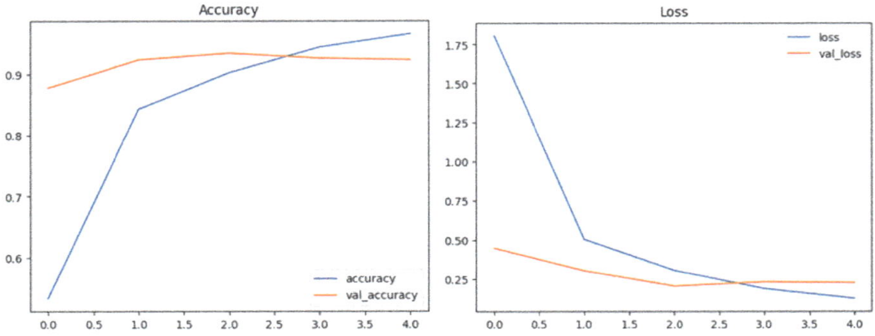

Fig. 6 Accurary and loss during training and validation process (x-axis: number of epochs)

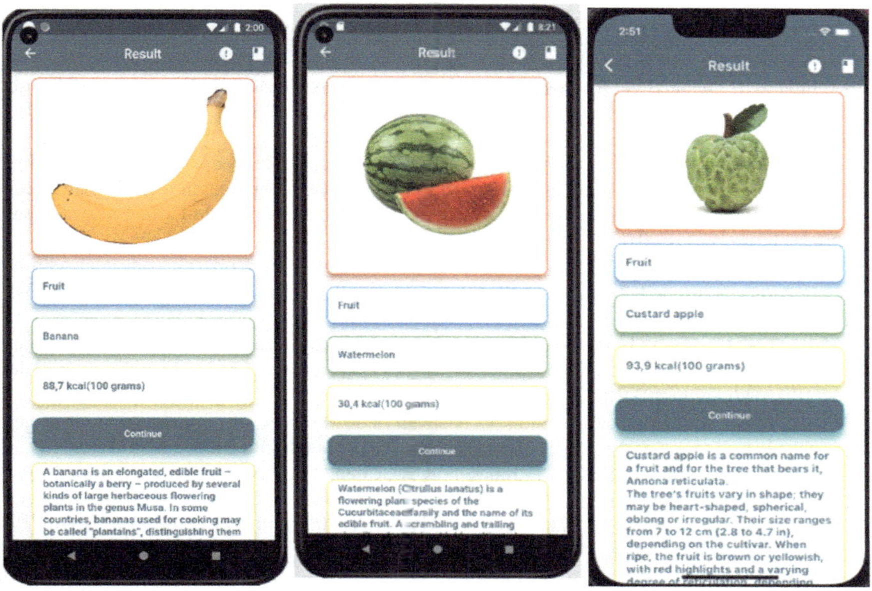

Fig. 7 Fruit calorie determination system on mobile devices

or the album for recognition. The interface includes two button. The first button is "Select a photo from the Camera" for choosing a fruit photo from the camera of mobile device. When you press the "Select a photo from the Camera" button, the application will move to the camera function on the phone device, where you can take a picture of the fruit you want and then press continue the system will return the results. The second button is "Select a photo from the Album" for choosing a photo fruit from the album of mobile devices.

5 Conclusion and Future Works

In this work, we propose a calorie determination system using deep learning approach, especially for dieters and athletes. This work firstly focuses on fruit calorie determination. First we collect the dataset of popular fruits in Vietnam and other public sources, then we propose using deep learning with MobilenetV2 architecture for fruit classification. An application is built for automatically deter-mining the calorie of those fruits from photo taken by smartphone. The method can achieve more than 90% accuracy and the application can work on both Android and IOS devices.

This work firstly focuses on fruit calorie determination, we continue to work on other kind of food such as meat, milk, egg, etc.

References

1. Mimma, N.E., Ahmed, S., Rahman, T., Khan R.: Fruits classification and detection application using deep learning. Sci. Program. (Article ID 4194874), 16 (2022). https://doi.org/10.1155/2022/4194874
2. Femling, F., Olsson, A., Alonso-Fernandez, F.: Fruit and vegetable identification using machine learning for retail applications. In: 2018 14th International Conference on Signal-Image Technology and Internet-Based Systems (SITIS), pp. 9–15. Las Palmas de Gran Canaria, Spain (2018). https://doi.org/10.1109/SITIS.2018.00013
3. Gill, H.S., Murugesan, G., Khehra, B.S., et al.: Fruit recognition from images using deep learning applications. Multimed. Tools Appl. **81**, 33269–33290 (2022). https://doi.org/10.1007/s11042-022-12868-2
4. Indira, D.N., Goddu, J., Indraja, B., Challa, V.M., Manasa, B.: A review on fruit recognition and feature evaluation using CNN. Mater. Today: Proc. **80**(Part 3), 3438–3443 (2023). https://doi.org/10.1016/j.matpr.2021.07.267
5. Vishnu, H.S., Sindhushree, B., Punith, A., Aishwarya, K., Praveen, G.: Fruit recognition system for calorie management. Int. Res. J. Eng. Technol. (IRJET) **08**(05) (2020)
6. Sandler, M., Howard, A., Zhu, M., Zhmoginov, A., Chen, L.-C.: MobileNetV2: inverted residuals and linear bottlenecks. In: 2018 IEEE/CVF Conference on Computer Vision and Pattern Recognition, pp. 4510–4520. Salt Lake City, UT, USA (2018). https://doi.org/10.1109/CVPR.2018.00474
7. Van-Quoc, V., Thai-Nghe, N.: Skin diseases detection with transfer learning. In: Saraswat, M., Chowdhury, C., Kumar Mandal, C., Gandomi, A.H. (eds.) Proceedings of International Conference on Data Science and Applications. Lecture Notes in Networks and Systems, vol. 551. Springer, Singapore (2023). https://doi.org/10.1007/978-981-19-6631-6_11
8. Thai-Nghe, N., Tri, N.T., Hoa, N.H.: Deep learning for rice leaf disease detection in smart agriculture. In: Dang, N.H.T., Zhang, YD., Tavares, J.M.R.S., Chen, BH. (eds.) Artificial Intelligence in Data and Big Data Processing. ICABDE 2021. Lecture Notes on Data Engineering and Communications Technologies, vol. 124. Springer, Cham (2022). https://doi.org/10.1007/978-3-030-97610-1_52

An Approach to Instrumental Song Classification Utilizing Spectrogram and Convolutional Neural Networks

Anh Tuan Le, Hien Thanh Thi Nguyen, Hoa Huu Nguyen, and Hai Thanh Nguyen

Abstract Searching for a song is a necessity, where the copyright of the song is a significant concern. This study proposes a method to classify and identify songs based on specific features that the model learns from music data. Python and CNN programming languages are used to build the model. In the first process, support libraries are used to extract audio data from the computer in WAV format. The dataset ***A*** includes 100 songs without lyrics, while the dataset ***B*** includes 100 audio files with the same song name but played in different types of musical instruments. We randomly cut the original audio files into clips less than 10 s long because users often use a specific code to find the entire track. The original audio files are split into clips of different lengths in the training set, including 1, 3, 5, 10, 20, 30, 60, and 90 s. Next, the Short-Time Fourier Transform was used to convert the audio data to the frequency domain. Finally, a shallow Convolutional Neural Network (CNN) and a Fully Connected layer (FC) were used to perform song classification tasks. We found that data augmentation by dividing the entire song into small pieces based on length significantly improved classification performance compared to those not using this technique. This research positively contributes to the advancement of e-commerce music systems, where listeners can enjoy music conveniently and memorably.

A. T. Le · H. T. T. Nguyen · H. H. Nguyen · H. T. Nguyen (✉)
Can Tho University, Can Tho, Vietnam
e-mail: nthai.cit@ctu.edu.vn

A. T. Le
e-mail: anhb1906362@student.ctu.edu.vn

H. T. T. Nguyen
e-mail: ntthien@ctu.edu.vn

H. H. Nguyen
e-mail: nhhoa@ctu.edu.vn

N. Hoang Phuong et al. (eds.), *Machine Learning and Other Soft Computing Techniques: Biomedical and Related Applications*, Studies in Systems, Decision and Control 543, https://doi.org/10.1007/978-3-031-63929-6_20

1 Introduction

Musical instruments have become an integral part of the daily lives of countless individuals worldwide. The meaning of musical instruments reflects people's diverse and rich musical preferences. According to data from Spotify platform [1], there are more than 10 million users across 7 European countries; they access a variety of music genres such as relaxing, Ambient, classical, electronic, jazz, etc. Through a mobile app or streaming, individuals can enjoy the instrument, which helps to have a tranquil environment, reduces stress, and fosters mental serenity. This experience improves mood and concentration during various activities, including work and study. Furthermore, musical instruments have universal appeal, overcoming language barriers and allowing people from different cultures to connect with music on an emotional level. Therefore, there are many demands to find a song from a short piece of music that people accidentally listen to from somewhere. Besides, the music industry's problem of copying and copyright infringement is enormous and familiar today.

With the development of technology and the internet, copying, distributing, or illegally using musical works has become more accessible and more popular than ever. The challenge is defining a song from its short audio excerpt. This study will research solutions to this challenge. The CNN [2] are used to analyze spectral images obtained from audio data, allowing users to explore diverse musical genres based on their acoustic characteristics. This approach lets users quickly and easily discover their favorite tracks, enhancing their music discovery journey. The critical contribution of this study consists of two important phases. Firstly, the proposed method randomly cuts original audio files into audio segments of less than 10 s because users often use a specific small segment to find the entire track. Secondly, during the training period, the original audio files were divided into audio segments of different lengths, including 1, 3, 5, 10, 20, 30, 60 and 90 s. This method determines which segments will provide the most accurate detection performance when applied to the recognition model. Besides, this study performs the classification and identifying process using machine learning models, mainly CNN, on spectrograms [3]. This approach highlights the study's uniqueness and demonstrates the specificity of using CNN to classify and search for specific acoustic properties of instrumental music. The following are our main contributions:

- We proposed an approach of using a shallow CNN network that can classify the song based on the spectrogram of the segment of the song. We also conducted to compare the experimental results related to data fragmentation and the effect of training on audio segments and the whole music; we found that when we break down a piece of music into smaller segments and train each part, classification results on segments are better than on the whole music.
- We proposed that data augmentation by dividing the songs into multiple segments could improve the classification accuracy and that CNN outperforms FC on both datasets. We found that performing data augmentation by dividing the dataset into segments improved the classification performance more than without applying this technique. This approach has significantly increased the performance of the

classification model. The data enhancement consists of dataset A, and additional data is added as fragments of different lengths, helping the model focus on learning and classification based on the unique characteristics of each part of the data in each song.
- In addition, we did a study to determine the importance of choosing a test song with a small number of seconds. The results show that choosing a shorter test segment improved the accuracy of the classification process. This can be explained by the fact that shorter segments often contain only a tiny portion of the music, focusing on the unique characteristics of specific instruments.
- Both CNN and FC models are applied to this study. By comparison, a vital exploration found that the CNN has a higher potential aggregation and performs better in the test phase.

This article is structured in the following order. In Sect. 2, we will first get acquainted with the research related to the topic. Then, in Sect. 3, we will investigate how to unravel the problem. Next, Sect. 4 Introduction to data used in research, data characteristics, and division, environmental setups. Finally, in Sect. 5, we will present our conclusions based on the experimental results.

2 Related Work

At present, there have been many remarkable studies with increasingly improved algorithms, bringing high Efficiency in finding instrumental music. These studies have achieved high accuracy and attracted the interest of music lovers worldwide, especially in countries with a long history of music with various instruments. Diverse traditions are imbued with ethnic cultures such as China, Thailand, India, and Mexico. The work in [4] was essential to the Search Engine Case Study. This study used approximately nearest neighbors to preprocess instrumental songs and extracted the characteristics of the tracks in the archive using Mel frequency Cepstral Factor (MFCC) extraction. The study resulted in 100 tracks of different lengths, with a sampling frequency of 16000, and each MFCC having a length of 13, yielding the best results. The accuracy in the Top 1 was 36%, the Top 5 was 4%, and the Top 10 was 44%. Additionally, the study applied the ANN algorithm and MFCC feature extraction to perform song search by harmony, helping to select the most suitable parameters for the most accurate results and the fastest search time and built a website that integrated algorithms to help users find songs more easily. The work in [5] was an essential work in the field of environmental sound recognition using deep neural network models, especially CNN. This article evaluated the potential of convolutional neural networks in classifying short audio clips of environmental sounds. A plunging model consisted of 2 convolutional layers with maximum pooling and two fully connected layers trained on a low-level representation of audio data (segmented spectrogram) with delta. The network's accuracy was evaluated on three publicly available datasets of urban and environmental records. This model outperformed baseline implementations based on mel frequency cepstral coefficients and achieved results comparable to other state-of-the-art approaches.

The work in [6] was essential to musical information extraction and automatic note recognition. The study focused on converting an audio signal into a frequency-time (spectral) representation and using a machine-learning model to predict musical notes and corresponding musical instruments to develop a system. Systems automatically extracted information from audio signals, such as note names and specific instruments in polyphonic music. They used pitch and fundamental frequency features to describe audio characteristics in expressions with spectrum maps. The study employed machine learning models such as Support Vector Machines (SVM) to classify musical notes and instruments. Models were trained on musical datasets annotated with musical notes. The authors tested the music dataset and compared the predicted results with the label information. The results showed the feasibility and potential of the method in realizing musical notes and instruments from audio signals. This led to the development of AMT using machine learning models, which opened the way to automating the processing of traditional music recordings.

The study [7] proposed a scheme to distinguish between vocal (song) and non-vocal (instrumental) music signals. They used similarity features based on spectral images to classify audio signals. Research observed that spectral images of musical instrument signals showed more stable peaks over time, and this was not the case for a song. It promoted research into finding features based on spectral images. Contextual features were calculated based on the occurrence pattern of the most critical frequencies over the time scale and the overall texture pattern revealed by the time-frequency distribution of the signal intensity. For classification, the authors relied on the Random Sample Consensus (RANSAC) technique that could handle many data types in one class. RANSAC created models for each class and performed classification model production based on it. The author prepared a music database consisting of 300 instrument files and 300 song files. Each file contained audio of approximately 40–45 s in duration. Files were obtained from burning CDs, recording live shows, and downloading from various sites on the Internet. The sampling of the given frequency data was 22050 Hz. Samples were 16-bit and monochrome. Data on flute, piano, guitar, and drums were available and saved. The songs also came in many genres like classical, jazz, rock, and Bhangra (also a North Indian genre). Audio files were divided into frames to calculate the characteristics. Corresponding to each frame was usually obtained. Each frame comprised 256 samples, including 128 cross-samples between two consecutive frames. Achieved results: Neural Network: instrument 85%, Song (80.33%), Overall (82.67%), RANSAC: instrument 97%, Song (93%), Overall (95%). In this study, the performance of RANSAC on the computed feature set was examined, and Neural Network performance on the same feature set showed that the performance of RANSAC was much better than that of Neural Network.

One of the notable studies was conducted by the authors of the study [8]. Unlike traditional methods, the research explored a classification scheme for musical instruments using features learned from a CNN. However, another method for feeding the phase information into a neural network was proposed in this research. Because a time series is a sequence of one-dimensional data, the filter (kernel) used for the time series in CNN should also be one-dimensional. This restriction gave rise to a limitation when applying a filter convolution because information in a specific temporal location could only be convoluted once. In contrast, a two-dimensional spectrogram image could be analyzed using a two-dimensional filter multiple times per single temporal location, providing more dimensions for analysis. The research initiative aimed at enhancing the performance of a classifier through data augmentation techniques. Specifically, the researchers shifted a time block by 13 samples, creating 13 temporally shifted Multiresolution Recurrence Plots (MRP) datasets. These datasets were then utilized to construct seven layers of RP images, each with a quadrupled length compared to the preceding block in the time series. The temporal sizes of these layers were set at 25, 27, 29, 211, 213, 215, and 217. By combining the proposed MRPs and spectrogram images with a multi-column network, they obtained an improved instrument classification performance using the UIOWA MIS database. The research also tested the performance of the combined network for a piano classification task. The results show that the classification of musical instrument timbre could be improved using MRP data with a spectrogram image and by feeding the data and image into multi-column CNN.

This study [9] had two primary contributions: first, it proposed a deep convolutional neural network architecture for environmental sound classification - model SB-CNN, PiczakCNN, SKM. Second, it proposed the use of audio data augmentation to overcome the problem of data scarcity and explored the influence of different augmentations on the performance of the proposed CNN architecture. The deep CNN architecture proposed in this study comprised three convolutional layers interleaved: 24, 48, and 48 filters with two pooling operations, followed by two fully connected (dense) layers. The research experimented with four different audio data augmentations (deformations): Time Stretching (TS), Pitch Shifting (PS1), Pitch Shifting (PS2), Dynamic Range Compression (DRC), and Background Noise (BG). Each deformation was applied directly to the audio signal before converting it into the input representation to train the network (log-mel-spectrogram). Combined with data augmentation, the proposed model achieved state-of-the-art results for environmental sound classification. The study improved performance by combining a deep, high-capacity model and a boosted training dataset. This combination outperformed the proposed CNN without boosting and the model "shallow" dictionary geometry with enhancement. Finally, the study examined the influence of each boost on the model's classification accuracy for each class and observed that each boost affected each class's accuracy differently, suggesting that the model performance could be further improved by applying conditional data augmentation to each layer.

Although many studies attempted to propose methods based on deep CNNs to perform song detection and data augmentation by shifting the time block, the efficiency of increasing the samples by separating the complete original songs into smaller pieces trained by a shallow CNN has not yet been clarified on instrumental songs. In addition, it also starts from the fact that we request the name of the song, but we usually only have a small piece of the song. Therefore, our study aims to evaluate the potential benefit of such an approach in improving the instrumental song classification tasks and support seeking a song with only a tiny piece of the song.

3 Methods

In the study, the strategy proposes a method to classify songs based on specific characteristics that the model learns from the music data. Python and CNN programming languages are used to build the model. In the first process, supporting libraries extract audio data from the computer in WAV format. The dataset ***A*** includes 100 songs without lyrics, while the dataset ***B*** includes 100 audio files with the same song name but played with different types of musical instruments. Then, the pydub library [10] randomly cuts the original audio files into clips less than 10 s long because users often use a specific snippet to find the entire track. The original audio files are split into clips of different lengths in the training set, including 1, 3, 5, 10, 20, 30, 60, and 90 s. Next, Short-Time Fourier Transform (STFT) is used to extract the musical features from the audio signal, converting them into spectral images. Next, the CNN model is built after having the optical image feature data spectrum. This neural network architecture is specifically designed to work with 2D data. In addition, using the CNN-1/FC models helps us to learn features at local locations on the image, thereby generating higher-level features for image classification. The workflow is shown in Fig. 1.

3.1 Data Collection and Division of Songs

A technical method to split an audio file into small segments of different duration, including 1, 3, 5, 10, 20, 30, 60, and 90 s. We used the Python pydub library. The process starts with choosing the length of the sub-audio tracks, which can be 1, 3, 5, 10, 20, 30, 60, or 90 s. Then, we go through each part of the original audio file. For each part, we calculate the start and end times. Check if the end time exceeds the length of the audio file, and if so, restrict it from exceeding it. Then, we create a name for the new sub-audio file by combining the original filename with the part number.

We use the calculated start and end times to perform the audio clipping from the original file and convert the child audio track into a new WAV file. The code then continues by setting the output directory path and looping through the audio files in the input directory. For each audio file, the code defines the input audio file path and the target directory for the sub. If the destination directory does not already exist, it will be created. Finally, split the audio file into several small parts, with each part

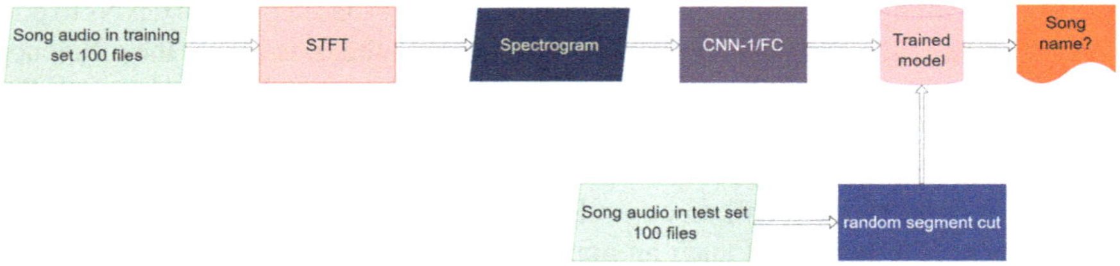

Fig. 1 Overall workflow for song classification with Spectrogram and machine learning algorithms

Table 1 Information on datasets used in the experiments. AXstest, BXstest includes samples where each sample is a piece with a length of X seconds randomly extracted from a song in dataset A, B, respectively, while AYstrain consists of sub-sequences of audio files with the length of Y seconds extracted from A

Datasets	The number of samples
A	100
B	100
A10stest	100
B10stest	100
B05stest	100
A01strain	26122
A03strain	8738
A05strain	5265
A10strain	2660
A20strain	1353
A30strain	919
A60strain	481
A90strain	346

length, and save it in a folder. A technical method used to split up audio files randomly to custom lengths is to use the pydub library in Python. This process begins with the total time of the original audio file. Then, randomly select the start time to cut to generate random audio parts. Once the audio sections have been created, we create an output directory based on the name of the original audio file to store the split audio tracks. These audio tracks are exported in WAV format and are automatically saved to the output folder. The result is a directory of custom-length random audio tracks from the original audio file.

In Table 1, dataset ***A*** includes 100 songs without lyrics, while dataset ***B*** includes 100 audio files with the exact song title but played with different instruments. This brings variety and richness to our data sets to study and analyze the influence of instruments and duration distribution on music. AXstest and BXstest include samples where each sample is a piece with a length of X seconds randomly extracted from a song in datasets A and B, respectively. At the same time, AYstrain consists of sub-sequences of audio files with the length of Y seconds extracted from A. As shown in Table 1, for example, A03strain contains audio segments cut from dataset A, having the same length of three seconds. Similarly, A01strain, A05strain, A10strain, A30strain, A60strain, and A90strain include samples with lengths of 1, 5, 10, 30, 60, and 90 s, respectively.

3.2 *Transforming Audio Signal to Image with Spectrogram*

In our research on spectrogram finding and classifying instrumental music based on the method of the STFT, an essential technique in signal and audio processing. The audio signal from the files is divided into small frames with a fixed length. This division helps determine the frequency variation over time in each moment.

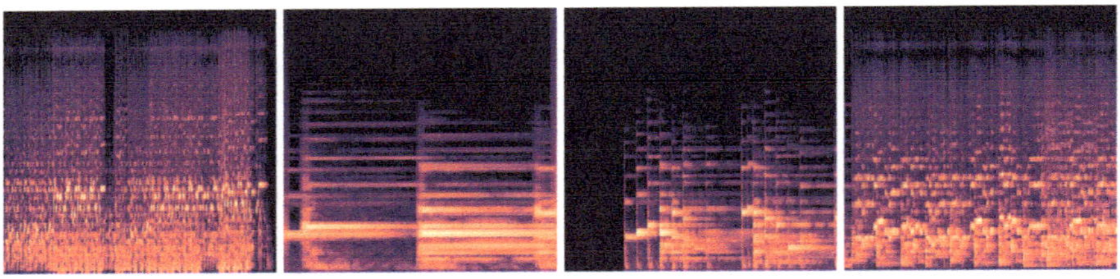

(a) The original. (b) one seconds (c) 10 seconds (d) 30 seconds

Fig. 2 Some illustrations of spectral images with various lengths cut from a song

Apply the Discrete Fourier Transform (DFT) to the audio signal in each frame. DFT decomposes the signal into individual frequency components and displays the magnitude of each frequency. The result of the DFT in each frame is a sequence of frequency-intensity values. The result is a 2D histogram of the different frequency bands in each timeframe. Spectroscopy allows us to observe changes in the frequency of sounds over time and detect musical characteristics such as tone, instrumentation, and noise changes. In the image, the time axis represents the track's duration, and the frequency axis represents the range of audio frequencies. Colors and variations in the spectral image represent the level and distribution of audio frequencies at a particular time. This is a unique way of expressing music information with images, helping us capture and analyze the characteristics and sound elements of the song. Figure 2 is an illustration of the spectrum of the original song "Ai chung tinh duoc mai" and the individual audio cuts (a complete song) from the dataset into one-second, 10 and 30 s.

3.3 The Networks for Song Recognition

We selected the architecture illustrated in Fig. 3; we called the network CNN-1 (with one convolutional layer) for short. A shallow convolutional network architecture can work well on Synthetic Image Representations as revealed in [11] with the input size of 32×32. The images generated by Spectrogram in this study also have the size of 32×32, so we use a similar architecture with [11] to perform the song classification tasks. Figure 3 exhibits that the network receives three-channel color images (audio file visualized by Spectrogram) as the input. This image is passed through a series of convolutional layers with 64 kernels of size 3×3 (with the stride 1), then passed through a maximum 2×2 composite layer (stretch 2). The ReLU activation function is applied after each convolution layer to produce the output. Next, a maximum composite layer (MaxPooling2D) is added to the model. This layer performs up to 2×2 summation on the output of the previous convolutional layer. It helps reduce the output size and retain the essential features. The next layer is the flat layer (Flatten), which transforms the output from the previous layers into a 1D vector. Finally, a fully connected (Dense) class is added to the model. This class has units

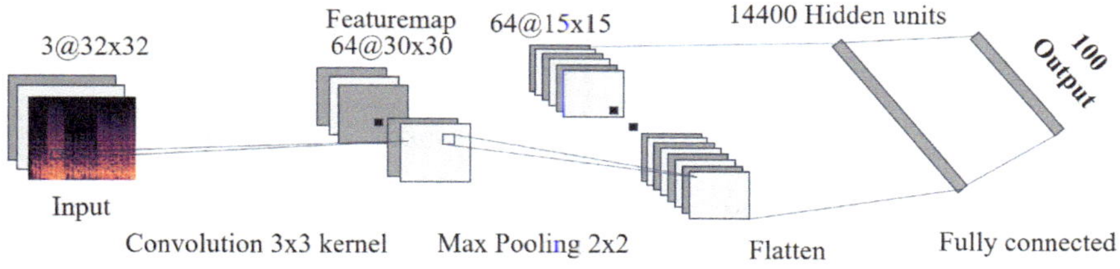

Fig. 3 The CNN-1 architecture includes a stack of one convolutional layer with 64 filters of 3×3 and a max pooling of 2×2, followed by one fully connected layer

= NUM-CLASSES, where NUM-CLASSES is 100 songs. The activation function used is 'softmax,' an activation function commonly used in classification problems to calculate the probability of each class given the input data. See the architecture of CNN-1 illustrated for the 32×32 image in Fig. 3 for more details. Besides CNN-1, we have also implemented a Fully Connected neural network (FC) that received the images and performed the linear regression to provide 100 outputs corresponding to 100 songs.

4 Experimental Results

4.1 Experimental Setup

All networks have used Adam for the optimizer, a batch size of 64, a learning rate of 0.0001, running to 500 epochs. CNN-1 has 1,441,892 Parameters, while FC owns 307,300. Several libraries are also implemented, such as annoy (version 1.17.0), librosa (version 0.9.2), python-speech-features (version 0.6), numpy (version 1.21.5), jar (version 1.1.2), tensorflow (version 2.8), and matplotlib (version 3.5.1).

4.2 The Length of Pieces Extracted from the Song Can Affect the Song Detection Performance

AXstest, BXstest includes samples where each sample is a piece with a length of Xseconds randomly extracted from a song in dataset A, B, respectively. Our idea was to test how many seconds of audio would best fit into the recognition model and be able to detect the song most accurately. In addition, we applied two methods, CNN-1 and FC models, to this dataset to compare the performance between the two network architectures in the task of instrumental music classification based on audio data. Furthermore, we highlighted the impressive results on the datasets to facilitate precise observation and comparison of the song classification performance between the CNN-1 and FC models based on Table 2.

Table 2 Efficiency of separation into small pieces in song classification tasks

	Models	CNN-1	FC	Models	CNN-1	FC	Models	CNN-1	FC
Roles	Datasets	Accuracy	Accuracy	Datasets	Accuracy	Accuracy	Datasets	Accuracy	Accuracy
Training set	A	1	1	A	1	0.9900	A	1	0.9400
Test set	A10stest	0.0900	0.0300	B10stest	0.0700	0.0300	B05stest	0.0500	0.0300
Training set	A01strain	**0.8318**	**0.6670**	A01strain	**0.9934**	**0.7598**	A01strain	**0.6236**	0.6805
Test set	A10stest	**0.7800**	**0.5200**	B10stest	**0.2100**	**0.1600**	B05stest	**0.2000**	0.1700
Training set	A03strain	0.9489	0.7906	A03strain	0.9205	0.7906	A03strain	0.9205	0.7906
Test set	A10stest	0.7000	0.4500	B10stest	0.2000	0.1300	B05stest	0.1800	0.1500
Training set	A05strain	0.9823	0.7504	A05strain	0.9853	0.7504	A05strain	0.9781	**0.7813**
Test set	A10stest	0.7200	0.4400	B10stest	0.2100	0.1500	B05stest	0.2000	**0.2000**
Training set	A10strain	0.9987	0.8680	A10strain	0.9857	0.701	A10strain	0.9751	0.7650
Test set	A10stest	0.5400	0.3200	B10stest	0.1900	0.1100	B05stest	0.1600	0.1000
Training set	A20strain	0.8847	0.8861	A20strain	0.9970	0.8580	A20strain	0.9741	0.7450
Test set	A10stest	0.4600	0.2200	B10stest	0.1500	0.1000	B05stest	0.1500	0.0900
Training set	A30strain	0.9028	0.9270	A30strain	0.9488	0.9140	A30strain	0.9630	0.8650
Test set	A10stest	0.2800	0.1400	B10stest	0.1400	0.0700	B05stest	0.1100	0.0700
Training set	A60strain	0.9563	0.5571	A60strain	0.9708	0.4636	A60strain	1	0.5051
Test set	A10stest	0.2200	0.0700	B10stest	0.1200	0.0500	B05stest	0.1200	0.0600
Training set	A90strain	0.9826	0.6184	A90strain	1	0.5028	A90strain	0.9971	0.9219
Test set	A10stest	0.1400	0.0600	B10stest	0.0800	0.0600	B05stest	0.0800	0.0600

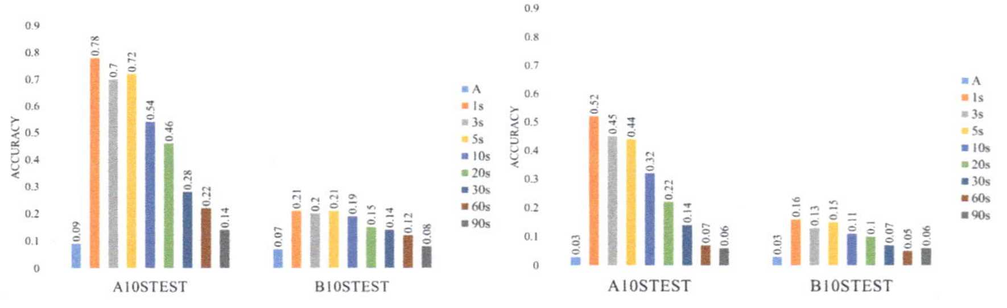

(a) CNN-1 (b) FC

Fig. 4 Comparison of FC and CNN-1 model performance on A10stest and B10stest datasets

Figure 4 presents a performance comparison between the two network structures CNN-1 and FC in classifying instrumental music in Table 2. Audio data is presented as cluster cells. The variety of test suites has yielded outstanding results for comparison and analysis. Both CNN-1 and FC models show high performance on data set with 1 s music, *A10stest* FC reaches 52%, CNN-1 model gives better results at 78%, *B10stest* FC achieves 16%, CNN-1 model gives better results at 21%. This impression is essential in realistic sound classification simulations since, in practice, listeners can start listening to any part of the track at random. Besides, the CNN-1 model still shows superiority over FC; CNN-1's ability to generalize and process complex data makes it a robust instrumental music classifier.

4.3 Data Augmentation on Songs

Table 3 evaluates the performance of dataset A in both cases: one is when no data augmentation is applied, and the other is when we did not add more data for the training set. AXstest where each sample is a piece with a length of Xseconds randomly extracted from a song in dataset A, respectively. During training, we built the training dataset by augmenting the "A" set data with audio clips of different lengths: 1, 3, 5, 10, 20, 30, 60, and 90 s. We also performed modeling of both CNN-1 and

Table 3 Song classification performance using and without data augmentation

				CNN-1		FC	
Roles	Datasets	Increased samples	Total	Accuracy	Time (seconds)	Accuracy	Time (seconds)
Without data augmentation							
Training set	A		100	1	14.1267	1	10.8605
Test set	A10stest		100	0.0900		0.0300	
Using data augmentation							
Training set	A + A01strain	26122	26222	**0.8312**	10091.3252	**0.6822**	360.4965
Test set	A10stest		100	**0.8000**		**0.5200**	
Training set	A + A03strain	8738	8838	0.9729	1679.7152	0.8176	213.0935
Test set	A10stest		100	0.6900		0.4800	
Training set	A + A05strain	5265	5365	0.9852	1371.6816	0.7940	109.1950
Test set	A10stest		100	0.7300		0.5000	
Training set	A + A10strain	2660	2760	0.9880	830.8003	0.8000	42.6509
Test set	A10stest		100	0.5500		0.3000	
Training set	A + A20strain	1353	1453	0.9587	477.6910	0.8066	28.0938
Test set	A10stest		100	0.4700		0.2200	
Training set	A + A30strain	919	1019	0.9754	177.2855	0.7595	26.2689
Test set	A10stest		100	0.3300		0.1300	
Training set	A + A60strain	481	581	0.9552	125.3534	0.7796	14.2146
Test set	A10stest		100	0.2600		0.0800	
Training set	A + A90strain	346	446	0.9910	97.6455	0.9192	20.5288
Test set	A10stest		100	0.1700		0.0900	

FC networks on the advanced dataset to compare the performance between the two network architectures in the task of instrumental music classification based on audio data. For example, as in Table 3, *A03strain+A* contains audio-based substrings of the same length of 3 s, separate from the added *A* dataset, original dataset A. Similarly, *A01strain+A*, *A05strain+A*, *A10strain+A*, *A20strain+A*, *A30strain+A*, *A60strain +A*, *A90strain+A* consists of a data-enhanced original dataset A, samples of length 1, 5, 10,20, 30, 60, 90 s, respectively. Also, in the Advanced Sample column, we calculate the percentage of sample gain and how much % in the tested dataset; we highlighted the impressive results on the datasets to facilitate precise observation and comparison of the song classification performance between the CNN-1 and FC models.

4.4 Classification Algorithms Comparison

In this study, we conduct a detailed analysis of the performance of two models, FC and CNN-1, on audio datasets for instrument classification based on acoustic data. More specifically, in Table 2, the results of song detection performance from an excerpt randomly cut from 100 songs in the original data set "A" of a 10 s length FC model have an accuracy of only 52%. In contrast, the CNN-1 model has an accuracy of 78%. In addition, the song detection performance results from an excerpt randomly cut from 100 songs in the original data set "B" with a length of 10 s. The CNN-1 model has an accuracy of 21%, compared to the FC model, only 16%. In addition, the song detection performance results from an excerpt randomly cut from 100 songs in the original dataset "B" with a length of 5 s. CNN-1 model achieved 20% accuracy when the FC model only reached 17%. In addition to the above test results, we performed performance evaluation on both cases: no data enhancement and data enhancement are applied as presented in Table 3. In addition, we model both CNN-1 and FC networks on the advanced dataset to compare the performance between the two network architectures in the task of Classification of Instruments based on acoustic data. The exciting result is that the FC model only gives 3% accuracy without data enhancement, while the CNN-1 model achieves 9%. However, when applying data enhancement, the performance of the FC model improves significantly to 52%, but the CNN-1 model still outperforms with an impressive accuracy of 80%. The research results clearly show the superiority of the CNN-1 model over the FC model in the classification task based on audio data. This is proven through various tests on different data sets.

5 Conclusion

We introduced an approach involving data fragmentation based on the power of convolutional neural networks on images to perform song recognition. The results showed that data augmentation by dividing the dataset into small chunks based on length significantly improved classification performance compared to not using this technique. This contributed significantly to the tool classification process and produced a significant increase in the performance of the classification model. In addition to the above studies, we have presented test results on both CNN-1 and FC models, providing a detailed view of the comparison between the two network architectures in the task of instrument classification based on audio data. We found that FC had a lower synthesis potential and resulted in worse performance than CNN-1 on the test set, which is an important finding. The scalability of the proposed approach is a significant concern. With the potential for the number of classes to change as new songs are added, the paper should address how the model can adapt to this scenario. Future work could explore solutions to this scalability challenge.

References

1. Goldmann, M., Kreitz, G.: Measurements on the spotify peer-assisted music-on-demand streaming system. In: 2011 IEEE International Conference on Peer-to-Peer Computing, pp. 206–211 (2011)
2. LeCun, Y., Bottou, L., Bengio, Y., Haffner, P.: Gradient-based learning applied to document recognition. Proc. IEEE **86**(11), 2278–2324 (1998)
3. Melchert, O., Roth, B., Morgner, U., Demircan, A.: OptFROG-analytic signal spectrograms with optimized time-frequency resolution. SoftwareX **10**, 100275 (2019)
4. Nguyen, H.T., Vo, L.D., Tran, T.T.: Approximate nearest neighbour-based index tree: a case study for instrumental music search. Appl. Comput. Syst. **28**(1), 156–162 (2023). https://doi.org/10.2478/acss-2023-0015. Jun
5. Piczak, K.J.: Environmental sound classification with convolutional neural networks. In: 2015 IEEE 25th International Workshop on Machine Learning for Signal Processing (MLSP), pp. 1–6. IEEE (2015)
6. Benetos, E., Dixon, S., Giannoulis, D., Kirchhoff, H., Klapuri, A.: Automatic music transcription: breaking the glass ceiling (ISMIR 2012)
7. Ghosal, A., Chakraborty, R., Dhara, B.C., Saha, S.K.: Song/instrumental classification using spectrogram based contextual features. In: Proceedings of the CUBE International Information Technology Conference, pp. 21–25 (2012)
8. Park, T., Lee, T.: Musical instrument sound classification with deep convolutional neural network using feature fusion approach (2015). arXiv:1512.07370
9. Salamon, J., Bello, J.P.: Deep convolutional neural networks and data augmentation for environmental sound classification. IEEE Signal Process. Lett. **24**(3), 279–283 (2017)
10. Boteju, W., Herath, H., Peiris, M., Wathsala, A., Samarasinghe, P., Weerasinghe, L.: Deep learning based dog behavioural monitoring system. In: 2020 3rd International Conference on Intelligent Sustainable Systems (ICISS), pp. 82–87. IEEE (2020)
11. Nguyen, T.H., Prifti, E., Sokolovska, N., Zucker, J.D.: Disease prediction using synthetic image representations of metagenomic data and convolutional neural networks. In: 2019 IEEE-RIVF International Conference on Computing and Communication Technologies (RIVF). IEEE (2019). https://doi.org/10.1109/rivf.2019.8713670

Heterogeneous Transfer Learning Using Pre-trained Feature Mapping and Exchange

Nguyen Thuan Duong, Dinh Thang Duong, Hong Phuc Nguyen, and Quang Vinh Dinh

Abstract Transfer Learning (TL) is an efficient approach for training neural networks with improved performance and lower costs compared to training from scratch. However, traditional TL techniques have drawbacks in mapping knowledge to the parameters of new networks, as they must necessarily use the same backbone as pre-trained models. To overcome this challenge, we introduce TLVFC, a novel method that enables knowledge transfer between networks with different architectures. For convolution layers, target model weights are initialized by identifying suitable pre-trained candidate layers and adjusting their parameters through variance to achieve alignment. For fully-connected layers, a new weight initialization approach is implemented that utilizes the distribution of pre-trained weights. To make the most of the model trained on a large dataset, an innovative feature exchange technique is further applied in the fine-tuning pipeline. Our method demonstrates remarkable convergence speed, outperforming the current state-of-the-art DPIAT [1] by over 2% in accuracy across various image classification benchmarks. The source code is available at https://github.com/ThuanNaN/TLVFC.

N. T. Duong · D. T. Duong
Department of Computer Science, Vietnam National University - University of Information Technology, Ho Chi Minh City, Vietnam
e-mail: 19522312@gm.uit.edu.vn

D. T. Duong
e-mail: 19522195@gm.uit.edu.vn

H. P. Nguyen
Department of Software Engineering, Eastern International University, Binh Duong, Vietnam
e-mail: phuc.nguyenhong@eiu.edu.vn

Q. V. Dinh (✉)
Department of Computer Science, Vietnamese German University, Binh Duong, Vietnam
e-mail: vinh.dq2@vgu.edu.vn

N. Hoang Phuong et al. (eds.), *Machine Learning and Other Soft Computing Techniques: Biomedical and Related Applications*, Studies in Systems, Decision and Control 543, https://doi.org/10.1007/978-3-031-63929-6_21

1 Introduction

In the ever-expanding domain of deep learning, Transfer Learning (TL) has revolutionized how models are trained and deployed. By allowing the reuse of pre-trained models on new tasks, traditional TL offers an efficient way to build upon pre-existing knowledge, significantly reducing computational costs and training times. Leveraging previously learned parameters has enabled researchers to tap into the potential of large-scale and complex architectures. As a result, TL has become one of the most significant breakthroughs in modern machine learning.

However, despite its powerful capabilities, TL is not without limitations. A significant constraint is the inability to transfer weights between networks that differ in architecture. This challenge makes standard methods less effective and inflexible, especially in the context of various network designs that have been proposed without the release of pre-trained models. For instance, in the original architecture of VGGNet [2], the convolution layers with small kernel-sized (3×3) are stacked for computational optimization of GPU. Recently, some research shows that using a more extensive kernel-sized network comes up with a more efficient solution, such as Swin Transformers [3], ConvNeXt [4], ConvNeXt-V2 [5], etc. To adapt to this growing trend, new methods must be developed that build on the principles of traditional TL, fully exploiting its potential while embracing the diverse and evolving landscape of modern deep learning architectures.

In this paper, we introduce TLVFC, a novel method that enables more effective and robust knowledge transfer without any restrictions on network design. By taking advantage of the rich information within sizeable pre-trained source models, TLVFC presents a new frontier in utilizing big models, demonstrating remarkable convergence speed and accuracy improvements across various benchmarks. Our main contributions can be summarized as follows:

1. We propose TLVFC, a new method that effectively leverages weights between two CNN-based networks that differ in structure.
2. Our method achieves state-of-the-art results on different datasets, and we further evaluate the performance of some transfer learning methods through image classification tasks.

2 Related Works

From the beginning of neural network development, there has been a lot of interest in figuring out the best way to set up the initial weights for efficient training. The understanding that initial weight values significantly impact the training process and overall model performance has motivated numerous studies in this area [6, 7]. Based on various proposed methods, we separate the problem of initializing weight into four main approaches:

Random weight initialization: In the early days, weights were randomly generated following a uniform or Gaussian distribution. These naive attempts were soon challenged by exploding and vanishing gradient phenomena, leading to poor model convergence. Several proposed methods have addressed this problem, including [8–10], etc., which obtained significant results. Nonetheless, those still facing some disadvantages related to requiring a lot of iterations for better performance might only be beneficial on specific network designs.

Transfer learning with pre-trained models: It turns out that starting training with weights taken from models trained on a large dataset achieves not only incredible model performance but also faster convergence [6, 7, 11]. This approach has become the baseline for training new models, but it is limited to working only on target models with the same architecture as pre-trained models.

Heterogeneous Transfer Learning: Transfer learning traditionally applies knowledge from one domain to another. When the source and target domains share analogous feature spaces and tasks, it is categorized as homogeneous transfer learning [12, 13]. On the other hand, significant disparities in feature representations, tasks, or model architectures lead to heterogeneous transfer learning [6, 7, 14, 15]. A salient challenge here is inter-architecture knowledge transfer, especially among varied CNN models like VGG16 [2] and ResNet [16]. To solve this issue, DPIAT [1] offers weight initialization using any pre-trained models. While their approach yields promising results, using a fundamental tensor resizing technique to handle the discrepancy in shape among the weight tensors is not entirely efficient and rational. In light of this, our method presents a more suitable direction to tackle this challenge.

3 Proposed Method

Our research study introduces the TLVFC method, which combines a three-stage pipeline called the copy-weight process. This process facilitates effective layer weight transition from pre-trained weights on large datasets like ImageNet1K [17] to various CNN-based networks. Additionally, a specialized technique is integrated into the training forward stage to enhance the utilization of the source model and the performance of the target model (Fig. 1).

Theoretically, convolutional layers act as feature extractors that learn general information. Meanwhile, fully-connected (FC) layers work as a classifier that learns distinctive characteristics of some specific objects [18]. Therefore, the copy-weight process is separated into two parts. In the feature extractor part, our primary concentration is on investigating candidate layers within the source model. We strive to align these layers with their respective counterparts in the target model and adjust their shape. This ensures a perfect fit while retaining the essential knowledge acquired during the pre-training stage. In the classifier part, TLVFC makes use of the distribution of FC layer parameters in a pre-trained model to perform weight initialization.

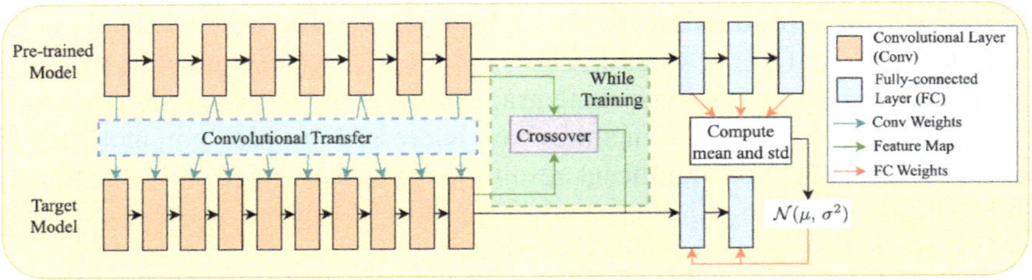

Fig. 1 Illustration of our proposed TLVFC method. In the pre-trained model, convolutional layers are matched and rectified to fit into the feature extractor of the target model. The mean and std of the classifier are calculated using fully-connected weights before they are fed into the weight initialization of target model counterparts. Finally, a special feature exchange technique is applied to improve the training outcome and make full use of the source model

While starting to train the target model resulting from the copy-weight process, we further use a new tactic called **feature exchange** to improve the final output of the feature extractor. This operation is inspired by the process of producing new offspring from two-parent individuals in a genetic algorithm.

3.1 Stage 1: Matching

When performing convolutional weights migration from one network to another, it most likely appears to experience an incompatibility in the number of layers between them. Therefore, the need for finding an effective layer index alignment is important. We address this difficulty by applying a simple index mapping method inspired by the process of mapping two integer ranges in model quantization [19]. This allows us to pair the index of transfer candidates between the target network T and pre-trained network P. To this end, for any layer ith in T, the pairing index of candidate i_C is calculated using the formula below:

$$i_C = round(\frac{n_c^P}{n_c^T} * i_T + z)$$

- n_c^P, n_c^T: number of convolution layers in pre-trained and target models.
- i_T: current index of convolution layer of the new model.
- z: zero-point of range value. By default, $z = 0$.

3.2 *Stage 2: Convolutional Transfer*

CNN-based models are well-known for their diversity in network design. One might differ from the other by adjusting the number of filters or kernel size of a single convolutional layer. For that reason, the quest to discover a superior method for aligning well-trained weights perfectly with their target counterparts becomes paramount.

Given two W^P, W^T as weights between any two consecutive learning layers that contain weights from pre-trained and target network, both with the shape formatting as (f_{out}, f_{in}, k, k) where:

- f_{out}: the depth of current convolution layer.
- f_{in}: the depth of previous convolution layer.
- k: the kernel size of convolution layer.

Our next task is to refactor the shape W^P and ensure that the shapes W^P and W^T are identical for use with the target model. We tackle the problem in two steps: kernel padding and weight transfer. As a result, the vast majority of pre-trained knowledge is transferred fast while retaining the precision of usual transfer learning performance.

3.2.1 Kernel Padding

At this step, we adjust the kernel size of W^P to fit the kernel size of W^T. Consider two kernels $K^P \in \mathbb{R}^{(k^P, k^P)}$, $K^T \in \mathbb{R}^{(k^T, k^T)}$ between a pre-trained network P and a target network T. The value ps, which is used for determining the padding size of each edge to adequately convert any kernel of size k^P to k^T, is first calculated by the following:

$$ps = \frac{k^T - k^P}{2}$$

By default, elements belonging to the ps range are set with a constant value equal to 0. Otherwise, is it filled with $K^P_{i,j}$:

$$K^T_{i,j} = \begin{cases} 0, & \text{if } ps < i, j < k^T - ps \\ K^P_{i,j}, & \text{Otherwise} \end{cases}$$

Intuitively, this operation centers that original knowledge K^P into the target kernels. As a result, the feature map produced by the pre-trained model is almost retrained, as is its performance. Besides, the experiment of which values to place in residual positions other than zero will be further discussed in Sect. 4.3.1.

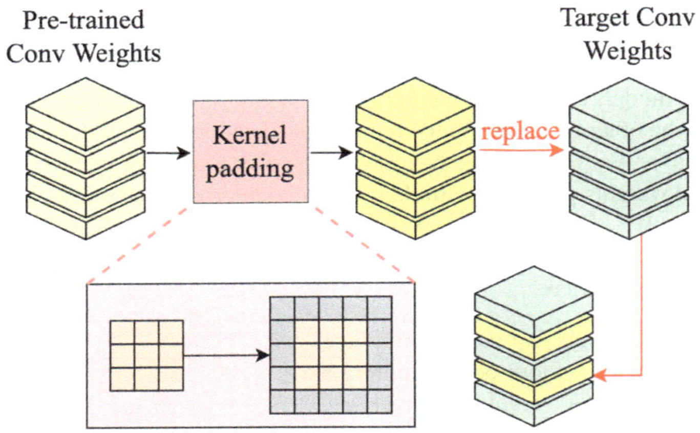

Fig. 2 Illustration of the convolutional transfer stage. The pre-trained weights are first padded to have a kernel size similar to the matched target. Then, knowledge transfer is processed using one of the three different operations depending on the shape of both weights

3.2.2 Weight Transfer

After dealing with the issue of different kernel sizes, we carry on the issue of copying weights that are distinct in shape (f_{out}, f_{in}) between pre-trained and target weights, respectively. The situation is likely to happen when the desired network configuration has any significant changes in the depth of any convolutional layers as opposed to the default settings in the model with prior training.

To properly make a weight transfer, the matched candidate weights are required to be in the same fashion by transforming into their opposite shapes. We categorize this step into three prominent cases, representing three differences in the number of filters our method might encounter. Assuming we have a pair of matched weights taken from a pre-trained network P and a target network T, the scenario might occur as follows (Fig. 2):

- $(f_{out}^P, f_{in}^P) > (f_{out}^T, f_{in}^T)$: In this circumstance, the size of pre-trained weights is larger than the size of target weights. Assuming each 4D tensor in a filter holds a specific feature, a better weight transition, in this case, would possess the ability to downsize the desired shape while still carrying crucial information from the list of source features. The **keep** technique is a way to reduce the shape of tensor weight. It is a selective process to choose specific weights. To determine which weights to keep, first, the variance lists for both the $W_{in}^P \in \mathbb{R}^{(f_{in},k_1,k_2)}$ and $W_{out}^P \in \mathbb{R}^{(f_{out},f_{in},k_1,k_2)}$ weights are calculated, representing each input and output filter of layer weight. These lists are then sorted, and the filters are retained by interlacing the position correlation of the sorted variance list to match the shape of the target weight (Fig. 4d). This approach ensures that the new weight is not influenced by biases from the original target or source weight (Fig. 5) and ultimately yields a high-performance outcome (Table 3).

- $(f_{out}^P, f_{in}^P) < (f_{out}^T, f_{in}^T)$: The laborious part comes at a state where weights from the target network require more space in-depth than the current capacity of the pre-trained model. To thoroughly restore the learned patterns, the transfer of all source weights to the target network is implemented. Based on the randomness idea of many weight initialization methods that involve generating values without any index prearrangement consideration, each of the pre-trained information is stochastically pasted into any possible position in opposition to one of the targets. The remaining parts are filled with the He initialization [8] values.
- $(f_{out}^P, f_{in}^P) = (f_{out}^T, f_{in}^T)$: When the shape of both selected weights is the same, the source weights are put into the target counterpart.

3.3 Stage 3: Fully-Connected Transfer

Classifiers in CNN-based neural networks are designed to learn distinct features for classification. Directly initializing target linear layers with pre-trained ones might not always be suitable. Instead, we propose a method that leverages pre-trained weights without direct copying. We calculate the mean and standard deviation (std) of each source linear weight and use the averaged values to generate the target linear weights. This positions the target model's classifier to learn object patterns effectively.

With the list of n_l^P linear weights in the pre-trained model W_i^P and n_l^t is the number of linear layers in the target network, the objective mean and std are found according to the following formulas:

$$clf\_mean_{avg} = \frac{1}{n_l^t n_l^P} \sum_{i=0}^{n_l^P} \left(\frac{1}{N_i} \sum_{j=0}^{N_i} W_{ij}^P \right)$$

$$clf\_std_{avg} = \frac{1}{n_l^t n_l^P} \sum_{i}^{n_l^P} \sqrt{\frac{1}{N_i - 1} \sum_{j=0}^{N-1} W_{ij}^P - \overline{W_i^P}}$$

To this end, every ith fully-connected weight of the target model is initialized by the below distribution:

$$W_i^T \sim \mathcal{N}(clf\_mean_{avg}, clf\_std_{avg}) .$$

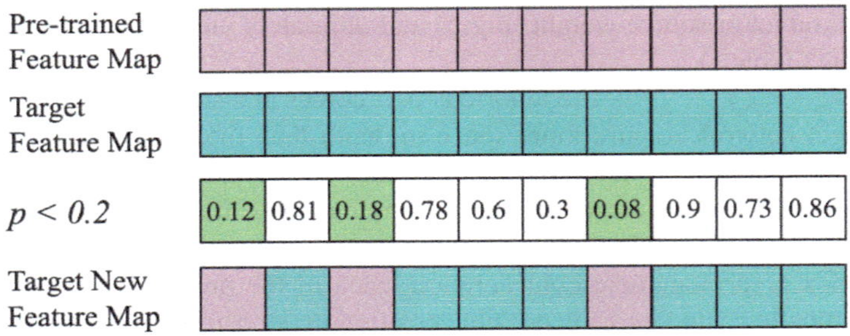

Fig. 3 Illustration of feature exchange. After flattening the final feature maps of pre-trained and target models, a random scalar is generated and compared for each target vector element against a threshold. This process determines whether the element is replaced with the corresponding pre-trained value

3.4 *Training with Feature Exchange*

To fully exploit the potential usage of a pre-trained model, we extend our work to customizing a traditional model training procedure to enhance the overall performance of the target models. Thoroughly, we tailor the forward stage by adding a so-called feature exchange to reproduce the last feature map of the target model. In this technique, the source and target model are used to extract the feature map at the last convolution layer, called vectors $V^t \in \mathbb{R}^{(batch, L^t)}$ and $V^p \in \mathbb{R}^{(batch, L^p)}$. Next, we employ a simple exchange algorithm between both output vectors, resulting in a new off-string feature map. This step acts as an augmentation technique that enriches the learned patterns, seeking to achieve a robust prediction. Last but not least, the exchange rate is set to $p_c < 0.2$ to avoid complexity in the exchange output, which might lead to confusion for the classifier. Following that, a number of $L^c = min(L^p, L^t)$ random probabilities $p_i \sim U[0, 1)$ $(p \in \mathbb{R}^{(batch, L^c)})$ are generated, and new element ith of target vector is calculated as below (Fig. 3):

$$V_i^t = \begin{cases} V_i^p, & \text{if } p_i < p_c \\ V_i^t, & \text{Otherwise} \end{cases} \quad i = 0 \ldots L^c$$

4 Experimental Results

4.1 *Setup*

We select one of the most popular tasks in the field, image classification, to evaluate the performance of our method.

Dataset. The evaluation is conducted on four datasets: CIFAR10, CIFAR100 [20], Food-101 [21], and PetImages [22]. The validation set is split from the training set with a ratio of 0.2, applying the same seed used in the training process.

Models. Experiments are executed with ResNet18 [16] as the target model and VGG16 [2] as the source model. Both models and the pre-trained ImagetNet1K weights are obtained in the Torchvision Python library.

Training. The same hyperparameter configuration is applied for all experiments, in which: Adam [23] is the optimizer with $\beta = (0.9, 0.999)$ and weight decay of 2e-6. A learning rate scheduler (STEPLR) with gamma = 0.1, beginning with 1e-4. The mini-batch size is 128 for CIFAR10, CIFAR100, and 64 for the PetImages and Food-101 datasets. Train within 200 epochs. The model with the best validation accuracy will be saved. The device that runs these experiments is the NVIDIA RTX 3090.

Metrics. We report all image classification tasks using the test accuracy score. Given a set of true images S and their corresponding predictions S^C. The accuracy is calculated by the following formula:

$$accuracy = \frac{S^C}{S}$$

4.2 *Using Cifar10, Cifar100, and PetImages Datasets*

We conduct a comparative analysis of the He initialization, DPIAT, and TLVFC across four datasets. We present three pipelines to highlight improvements in the performance of our method. First, the TLVFC-based approach focuses on weight transfer between the target and pre-trained models at the convolutional and fully connected layers. Next, the TLVFC-cross introduces the feature exchange method, generating a new feature map for the target model classifier. This approach significantly outperforms TLVFC-base on CIFAR100, Food-101, and PetImages. Finally, our comprehensive TLVFC combines all operations, achieving peak accuracy.

From Table 1, it can be observed that our proposed method achieves new state-of-the-art on four datasets compared to the current knowledge transfer DPIAT method. TLVFC outperforms DPIAT by over 2% in the Food-101 and PetImages datasets.

4.3 *Ablation Study*

4.3.1 Zero Padding

To expand the kernel size while preserving its feature extraction capabilities as in the original kernel, we position the core kernel at the center and initialize the surrounding padding. Table 2 shows that the best value for initializing padding is zero, which keeps the efficiency of the core kernel and does not break feature maps.

Table 1 Benchmarking results on four datasets. Methods are reported with top-1 accuracy. The target model is Resnet18, and the pre-trained model used for the transfer process is VGG16 with ImageNet1K weights

Method	CIFAR10	CIFAR100	Food-101	PetImages
He Init [8]	0.7635	0.4178	0.675	0.9481
DPIAT [1]	0.7695	0.4155	0.6806	0.9498
TLVFC-base	0.7677	0.4234	0.7038	0.9669
TLVFC-cross	0.7674	0.4293	**0.7043**	0.9686
TLVFC	**0.7724**	**0.4294**	0.7025	**0.9696**

Table 2 Performance of the padding method by He and Glorot initializers compared to zero padding on the VGG16 model with the kernel size increased to 5×5

Method		CIFAR10	NSD	PetImages
Initialization	He	0.7417	0.8556	0.9321
	Glorot	0.7662	0.8663	0.9347
Padding	He	0.8683	0.917	0.9835
	Glorot	0.8769	0.9213	0.9863
	Zero	**0.8845**	**0.9236**	**0.9893**

4.3.2 Copy Convolutional Weights Based on Variance

The technique of determining weights via variance effectively addresses the challenges in stage 2 of our proposed method. However, selecting the best among various variance strategies remains a challenge. Previous research on weight initialization for convolutional layers, such as He and Glorot, aimed to determine ideal starting distributions to prevent **exploding gradients** during feed-forward and **vanishing gradients** in backward propagation. Transferring weights through replication can change this distribution, potentially hindering the training process of the model.

Figure 4 simulates how different variance selection strategies applied in the copy-weight process work. Additionally, Fig. 5 shows the gradual shape transformation of the distribution of the weights in the target parameters. Those four approaches are described as follows:

- maxVar: Select weights by arranging all kernels in descending order of their variance values and choosing from the top. This approach tends to select kernels with the highest variance first, aligning closely with the original target weight distribution.
- minVar: In contrast to maxVar, minVar selects weights by arranging all kernels in ascending order of their variance values. This method prioritizes kernels with the lowest variance, selecting them first.
- twoTailed: Select weights starting from two tails of variance distribution.
- interLeaved: Selecting weights that interleave with each other regarding variance.

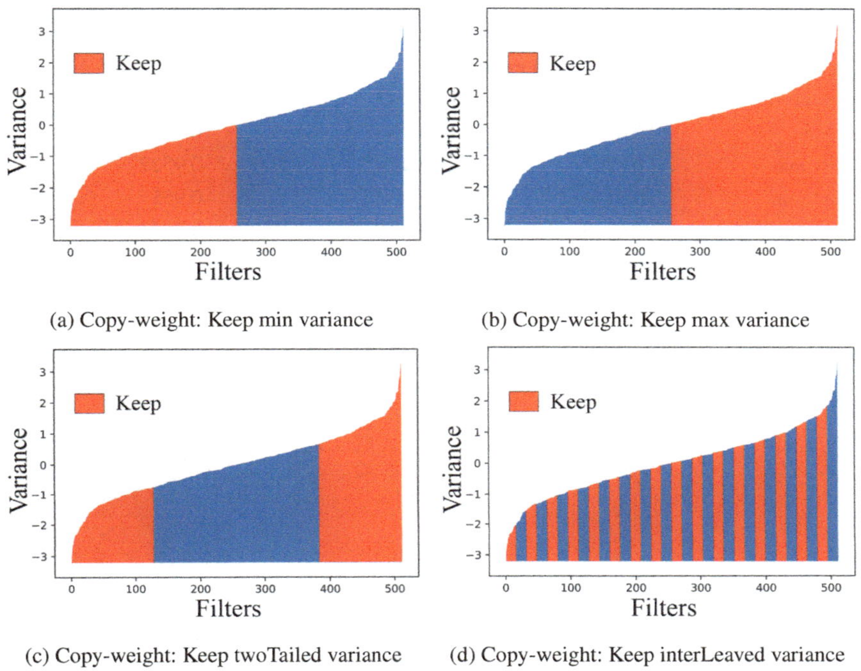

Fig. 4 The distribution of variance is sorted in ascending order for each layer weight. The selection method of weights is based on variance with different criteria: min, max, twoTailed, or interLeaved

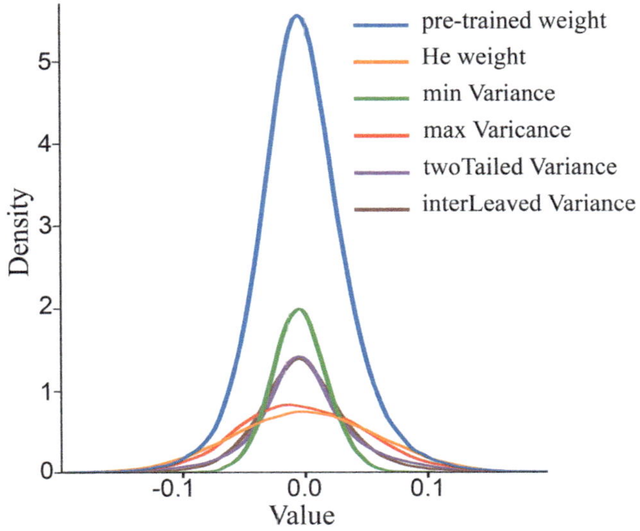

Fig. 5 The influence of the selection approach on the distribution of the weights. Apparently, twoTailed and interleaved selections show an average result among displayed distributions

Table 3 The performance of the experimental model (see Fig. 6) with the pre-trained model is VGG16 among weighted methods by retaining a portion of the weights according to specific criteria such as min, max, twoTailed, or interLeaved

Method	CIFAR10	NSD	PetImages
He Init	0.8021	0.871	0.9397
Keep: max	0.8207	0.88	0.9527
Keep: min	0.8174	0.8913	0.955
Keep: twoTailed	0.8209	0.888	0.9543
Keep: interLeaved	**0.8238**	**0.892**	**0.9569**

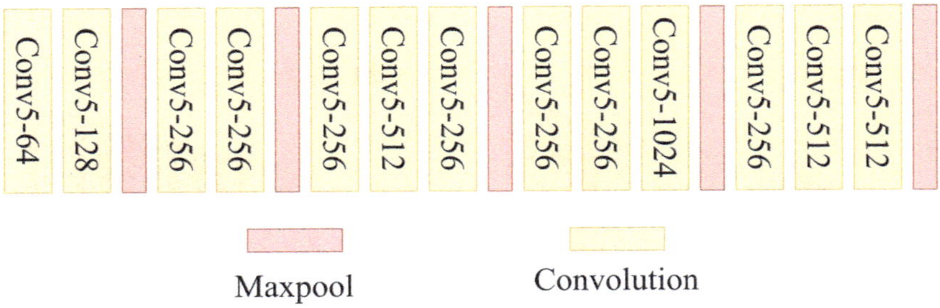

Fig. 6 The model is based on the design of VGG16. However, by modifying the output channel, the shape of the weight has been altered. Conv5 is the name given to the convolution layer after raising the kernel size to 5

Upon analyzing Fig. 5 and Table 3, it becomes evident that selecting weights solely based on the max and min variance is not particularly effective when compared to twoTailed and interLeaved.

4.3.3 Effectiveness on Very Deep Networks

TLVFC copies convolutional weights to the target network based on the index of each layer. When it comes to very deep networks, many objective layers will have the same weights as they share the same mapped index. To investigate the performance of TLFVC in this circumstance, an experiment with the Resnet series is conducted. The results displayed in Table 4 show us that our method yields a fruitful accuracy in Resnet with more layers than the least one.

Table 4 Performance of TLVFC on the Resnet family with pre-trained parameters from VGG19. As the number of layers increases, TLFVC achieves better results than the He initialization approach within 50 training epochs

	Resnet-34	Resnet-50	Resnet-101	Resnet-152
He Init	0.7524	0.705	0.6728	0.6435
TLVFC	0.7696	0.7483	0.7195	0.6979

5 Conclusion

In this paper, we presented a new mechanism to initialize weights and a new training process for a model trained from scratch. We first transferred knowledge from the pre-trained model to the equivalent mapped layer in the target model using weights at convolutional layers. Next, the fully connected layers of the target model were initialized by a normal distribution. The mean and standard deviation of this distribution are determined by the weight of pre-trained FC layers. Additionally, we leveraged a feature exchange strategy to enhance the performance of the target model during its training process. Through experiments conducted on various datasets, we showcased the remarkable performance and state-of-the-art results achieved by our proposed methodology.

References

1. Czyzewski, M.A., Nowak, D., Piechowiak, K.: Breaking the architecture barrier: a method for efficient knowledge transfer across networks (2022). arXiv:2212.13970
2. Simonyan, K., Zisserman, A.: Very deep convolutional networks for large-scale image recognition (2015). arXiv:1409.1556
3. Liu, Z., Lin, Y., Cao, Y., Hu, H., Wei, Y., Zhang, Z., Lin, S., Guo, B.: Swin transformer: hierarchical vision transformer using shifted windows. In: 2021 IEEE/CVF International Conference on Computer Vision (ICCV), pp. 9992–10002
4. Liu, Z., Mao, H., Wu, C.-Y., Feichtenhofer, C., Darrell, T., Xie, S.: A ConvNet for the 2020s. In: 2022 IEEE/CVF Conference on Computer Vision and Pattern Recognition (CVPR), pp. 11966–11976
5. Woo, S., Debnath, S., Hu, R., Chen, X., Liu, Z., Kweon, I.S., Xie, S.: ConvNeXt v2: Co-designing and scaling convnets with masked autoencoders. In: 2023 IEEE/CVF Conference on Computer Vision and Pattern Recognition (CVPR), pp. 16133–16142
6. Weiss, K., Khoshgoftaar, T.M., Wang, D.: A survey of transfer learning. J. Big Data **3**(9) (2016)
7. Zhuang, F., Qi, Z., Duan, K., Xi, D., Zhu, Y., Zhu, H., Xiong, H., He, Q.: A comprehensive survey on transfer learning. Proc. IEEE **109**(1), 43–76 (2021)
8. He, K., Zhang, X., Ren, S., Sun, J.: Delving deep into rectifiers: surpassing human-level performance on ImageNet Classification. In: 2015 IEEE International Conference on Computer Vision (ICCV), pp. 1026–1034
9. Glorot, X., Bengio, Y.: Understanding the difficulty of training deep feedforward neural networks. In: Proceedings of the 13th International Conference on Artificial Intelligence and Statistics (AISTATS), May 13–15, 2010, pp. 249–256

10. Nitta, T.: Weight initialization without local minima in deep nonlinear neural networks (2018). arXiv:1806.04884
11. Yu, F., Xiu, X., Li, Y.: A survey on deep transfer learning and beyond. Mathematics **10**(19), Article 3619 (2022)
12. Wang, W., Zhai, W., Cao, Y.: Deep Inhomogeneous Regularization For Transfer Learning. In: 2020 IEEE International Conference on Image Processing (ICIP), Abu Dhabi, United Arab Emirates, pp. 221–225 (2020)
13. Wei, S., Moore, R., Zhang, H., Xie, Y., Kamaleswaran, R.: Transfer causal learning: causal effect estimation with knowledge transfer. In: ICML 3rd Workshop on Interpretable Machine Learning in Healthcare (IMLH) (2023)
14. Wu, J., Wu, Y., Niu, N., et al.: MHCPDP: multi-source heterogeneous cross-project defect prediction via multi-source transfer learning and autoencoder. Software Qual. J. **29**, 405–430 (2021)
15. Bao, R., Sun, Y., Gao, Y., Wang, J., Yang, Q., Chen, H., Mao, Z.-H., Xie, X., Ye, Y.: A survey on heterogeneous transfer learning. J. Big Data **4** (2023)
16. He, K., Zhang, X., Ren, S., Sun, J.: Deep residual learning for image recognition. In: 2016 IEEE Conference on Computer Vision and Pattern Recognition (CVPR), pp. 770–778
17. Deng, J., Dong, W., Socher, R., Li, L.-J., Li, K., Fei-Fei, L.: ImageNet: a large-scale hierarchical image database. In: Proceedings of the 2009 IEEE Conference on Computer Vision and Pattern Recognition, pp. 248–255 (2009)
18. Jogin, M., Mohana, M., Madhulika, M., Divya, G., Meghana, R., Apoorva, S.: Feature extraction using convolution neural networks (CNN) and deep learning. In: 2018 3rd IEEE International Conference on Recent Trends in Electronics, Information and Communication Technology (RTEICT), pp. 2319–2323
19. Wu, H., Judd, P., Zhang, X., Isaev, M., Micikevicius, P.: Integer quantization for deep learning inference: Principles and empirical evaluation (2020). arXiv:2004.09602
20. Krizhevsky, A.: Learning multiple layers of features from tiny images. Technical Report TR-2009. University of Toronto, Toronto (2009)
21. Bossard, L., Guillaumin, M., Van Gool, L.: Food-101 - Mining discriminative components with random forests. In: Fleet, D., Pajdla, T., Schiele, B., Tuytelaars, T. (eds.) Computer vision - ECCV 2014. ECCV 2014. Lecture Notes in Computer Science, vol. 8694. Springer, Cham (2014)
22. Parkhi, O.M., Vedaldi, A., Zisserman, A., Jawahar, C.V.: Cats and dogs. In: Proceedings of the IEEE Computer Society Conference on Computer Vision and Pattern Recognition (CVPR), pp. 3498–3505 (2012)
23. Kingma, D.P., Ba, J.: Adam: a method for stochastic optimization (2017). arXiv:1412.6980

Usually, Either Left and Right Brains Are Equally Active or Only One of Them Is Active: First-Principles Explanation

Julio C. Urenda and Vladik Kreinovich

Abstract It is known that in most practical situations, either both left and right brains are equally active, or only one of them is active. A recent paper showed that this empirical phenomenon can be explained by a realistic model of the brain effectiveness. In this paper, we show that this conclusion can be made without any specific assumptions about the brain, based on first principles.

1 Formulation of the Problem

Empirical phenomenon that needs explaining. The brain of many living creatures—including, of course, humans—consists of two largely independent parts. One of the names for these two parts is left and right brain.

It turns out that in most practical situations:

- either left and right brains are equally active,
- or only one of them is active.

Specifically:

- when the corresponding task is not very complex, left and right brain are equally involved, but
- when the task is sufficiently complex, only one of the two brain halves is active.

How this phenomenon is usually explained. Several qualitative explanations of the above phenomenon have been proposed. Their main idea is that:

J. C. Urenda
Department of Mathematical Sciences, University of Texas at El Paso, 500 W. University, El Paso, TX 79968, USA
e-mail: jcurenda@utep.edu

V. Kreinovich (✉)
Department of Computer Science, University of Texas at El Paso, 500 W. University, El Paso, TX 79968, USA
e-mail: vladik@utep.edu

N. Hoang Phuong et al. (eds.), *Machine Learning and Other Soft Computing Techniques: Biomedical and Related Applications*, Studies in Systems, Decision and Control 543, https://doi.org/10.1007/978-3-031-63929-6_22

- on the one hand, the duplication coming from using both (reasonably independent) parts of the brain to solve a problem makes the result more reliable, but
- on the other hand, using both parts of the brain to solve a problem doubles the needed amount of energy; thus, when the task is complex and thus requires a large amount of energy, the duplication is no longer affordable.

A recent quantitative explanation. A recent paper [2] come up with a reasonable model $f(\ell, r)$ of how the brain's effectiveness depends on the activity levels ℓ and r of the left and right brains. This model includes some parameters that correspond to the complexity of the task. It turns out that in this model, for each level of complexity, the maximal effectiveness is attained:

- either when $\ell = r$
- or when one of the two activity levels is equal to 0.

In this model, in full accordance with the experimental data:

- for complex tasks, the maximal effectiveness is attained when one of the activity levels is 0,
- while for less complex tasks, the maximum effectiveness is attained when $\ell = r$.

Thus, this model full explains the empirical phenomenon.

A natural question. While the model proposed in [2] is reasonable, it provides—as any other model—only an approximate description of the real-life phenomenon. To more accurately describe the actual phenomenon, we most probably need a somewhat different model. It is therefore reasonable to ask whether the explanation from this paper can be extended to a more general class of models.

What we do in this paper. In this paper, we provide a partial answer to the above question. Namely, we show that in the natural quadratic approximation, every bounded model has the property that optimal effectiveness is attained:

- either when $\ell = r$
- or when one of the activity levels is equal to 0.

2 Definitions and the Main Result

We need bounded models. In practice, there is always a limit to our effectiveness. So, it is reasonable to only consider models $f(\ell, r)$ that are *bounded*:

Definition 1 We say that a function $f(\ell, r)$ is *bounded* if there exists a bound B for which $f(\ell, r) \le B$ for all $\ell \ge 0$ and $r \ge 0$.

We need realistic models.

- We know that the level of activities in each half of the brain affects the effectiveness, so we want the models that reflect this, i.e., for which $f(\ell, r)$ is not equal to the constant or to the function of only one of the two variables.
- We want to make sure that in the model, it makes sense for the brain to be active, i.e., at least some values of ℓ and r leads to better effectiveness than in situations when the brain is inactive, i.e., that we have $f(\ell, r) > 0$ for some ℓ and r.
- We also want to make sure that there is a difference between the left and the right brains, i.e., that the effectiveness is not simply the function of the total activation $\ell + r$.

Definition 2 We say that a function $f(\ell, r)$ is *realistic* if:

- it is not equal to a function of one variable,
- there exist values $\ell > 0$ and $r \geq 0$ for which $f(\ell, r) > f(0, 0)$, and
- it is not equal to a function of the sum $\ell + r$.

We will consider only symmetric models. In practice, left and right brain are largely similar. So, following [2], we will only consider models in which the dependence of effectiveness on the two activity levels is symmetric:

Definition 3 We say that a function $f(\ell, r)$ is *symmetric* if we have $f(\ell, r) = f(r, \ell)$ for all ℓ and r.

We will consider the simplest possible models. We do not know the exact shape of the dependence $f(\ell, r)$.

Such situations are ubiquitous in physics and in other application areas. In such situations, a usual practice is to consider a general form of the function—which, in the sufficient smooth cases, can be expanded in Taylor series—and keep only a few terms in this expansion; see, e.g., [1, 3].

The simplest case if when we only keep linear terms in this expansion, i.e., consider the dependence of the type

$$f(\ell, r) = a_0 + a_\ell \cdot \ell + a_r \cdot r \tag{1}$$

for some coefficient a_0, a_ℓ, and a_r. It turns out that in this approximation, we cannot have a function with the desired properties, even if we do not require symmetry:

Proposition 1 *No linear function is bounded and realistic.*

Comment. For reader's convenience, all the proofs are placed in the special Proofs section.

Since we cannot have linear model, the next simplest are quadratic models. A general quadratic function of two variables ℓ and r has the following form:

$$f(\ell, r) = a_0 + a_\ell \cdot \ell + a_r \cdot r + a_{\ell\ell} \cdot \ell^2 + a_{\ell r} \cdot \ell r + a_{rr} \cdot r^2 \tag{2}$$

For quadratic models, we have the following result.

Proposition 2 *For any bounded realistic symmetric quadratic function $f(\ell, r)$, its largest value is attained:*

- *either when one of the values ℓ and r is equal to 0,*
- *or when $\ell = r$.*

Discussion. Thus indeed, the above-described phenomenon follows from the first principles.

3 Proofs

Proof of Proposition 1.

1°. Depending on the signs of the coefficients a_ℓ and a_r, we can have the following situations:

- one of these two coefficients is positive;
- one of these coefficients is 0; or
- both coefficients are neither positive nor 0, i.e., both coefficients are negative.

Let us show that in all three cases, one of two desired conditions—that the function is bounded and realistic—is not satisfied.

2°. If at least one of the coefficients a_ℓ and a_r is positive, then the function $f(\ell, r)$ is not bounded.

3°. If at least one of these two coefficients is 0, then the effectiveness does not depend on the corresponding value, so the function is not realistic.

4°. If both coefficients are negative, then for $\ell \geq 0$ and $r \geq 0$, we cannot have $f(\ell, r) > f(0, 0)$, so the function is not realistic.

The proposition is proven.

Proof of Proposition 2. Let $f(\ell, r)$ be a bounded realistic symmetric quadratic function.

1°. Due to symmetry, we have $a_\ell = a_r$ and $a_{\ell\ell} = a_{rr}$. Thus, the expression (2) takes the following simplified form:

$$f(\ell, r) = a_0 + a_\ell \cdot (\ell + r) + a_{\ell\ell} \cdot (\ell^2 + r^2) + a_{\ell r} \cdot \ell \cdot r. \tag{3}$$

2°. Let us show that $a_{\ell\ell} \leq 0$.

Indeed, if this inequality was not true, i.e., if we had $a_{\ell\ell} > 0$, then for $r = 0$ and $\ell \to \infty$, the expression (3) would tend to infinity, thus the function $f(\ell, r)$ would not be bounded. Since we assumed that the function $f(\ell, r)$ is bounded, we have $a_{\ell\ell} \leq 0$.

3°. Let us now show, by contradiction, that we cannot have $a_{\ell\ell} = 0$, i.e., that we have $a_{\ell\ell} < 0$.

3.1°. Indeed, if $a_{\ell\ell} = 0$, the quadratic function has the form

$$f(\ell, r) = a_0 + a_\ell \cdot (\ell + r) + a_{\ell r} \cdot \ell \cdot r. \tag{4}$$

3.2°. Let us show that if $a_{\ell\ell} = 0$, then we must have $a_{\ell r} < 0$.
Indeed:

- we cannot have $a_{\ell r} = 0$, since then the function $f(\ell, r)$ will be linear, and this would contradict Proposition 1;
- we cannot have $a_{\ell r} > 0$, since then for $\ell = r \to \infty$, we would get $f(\ell, r) \to \infty$, and we assumed that the function $f(\ell, r)$ is bounded.

3.3°. Let us now show that if $a_{\ell\ell} = 0$, then we must have $a_\ell \leq 0$.
Indeed, if $a_\ell > 0$, then for $\ell \to \infty$ and $r \to 0$, we will have $f(\ell, r) \to \infty$, and we assume that the function $f(\ell, r)$ is bounded.

3.4°. We have shown that if $a_{\ell\ell} = 0$, then we have $a_\ell \leq 0$ and $a_{\ell r} < 0$. Thus, for $\ell \geq 0$ and $r \geq 0$, we have $f(\ell, r) \leq f(0, 0)$, and we assumed that the function $f(\ell, r)$ is realistic.

4°. We are interested in values (ℓ_0, r_0) for which the function $f(\ell, r)$ attains its maximum for $\ell \geq 0$ and $r \geq 0$. In particular, this means that:

- for $r = r_0$, the function $\ell \mapsto f(\ell, r_0)$ attains its maximum for $\ell = \ell_0$, and
- for $\ell = \ell_0$, the function $r \mapsto f(\ell_0, r)$ attains its maximum for $r = r_0$.

In general, the maximum of any smooth function on an interval is attained:

- either at one of the endpoints,
- or at a point where its derivative is equal to 0.

Here, the endpoints are $\ell = 0$ and $r = 0$. Thus, if the maximum is attained at one of these endpoint, for this case, the proposition is proven.

To complete the proof, we need to consider the case when both partial derivatives—with respect to ℓ and r—are equal to 0. In this case, we have the following two equalities:

$$\frac{\partial f}{\partial \ell} = a_\ell + 2a_{\ell\ell} \cdot \ell + a_{\ell r} \cdot r = 0; \tag{5}$$

$$\frac{\partial f}{\partial r} = a_\ell + 2a_{\ell\ell} \cdot r + a_{\ell r} \cdot \ell = 0. \tag{6}$$

Subtracting (6) from (5), we get

$$2a_{\ell\ell} \cdot (\ell - r) - a_{\ell r} \cdot (\ell - r) = 0,$$

i.e.,

$$(2a_{\ell\ell} - a_{\ell r}) \cdot (\ell - r) = 0. \tag{7}$$

5°. Let us show that we cannot have $2a_{\ell\ell} - a_{\ell r} = 0$.

Indeed, in this case, we would have $a_{\ell r} = 2a_{\ell\ell}$, so the function $f(\ell, r)$ would take the form

$$f(\ell, r) = a_0 + a_\ell \cdot (\ell + r) + a_{\ell\ell} \cdot (\ell + r)^2,$$

and we assumed that $f(\ell, r)$ is not a function of the sum $\ell + r$.

Thus, we cannot have $2a_{\ell\ell} - a_{\ell r} = 0$, so $2a_{\ell\ell} - a_{\ell r} \neq 0$. Dividing both sides of the equality (7) by this non-zero number, we conclude that $\ell - r = 0$, i.e., that $\ell = r$.

The proposition is proven.

Acknowledgements This work was supported in part by the National Science Foundation grants 1623190 (A Model of Change for Preparing a New Generation for Professional Practice in Computer Science), HRD-1834620 and HRD-2034030 (CAHSI Includes), EAR-2225395 (Center for Collective Impact in Earthquake Science C-CIES), and by the AT&T Fellowship in Information Technology.

It was also supported by the program of the development of the Scientific-Educational Mathematical Center of Volga Federal District No. 075-02-2020-1478, by a grant from the Hungarian National Research, Development and Innovation Office (NRDI), and by the Center of Excellence in Econometrics, Faculty of Economics, Chiang Mai University, Thailand.

References

1. Feynman, R., Leighton, R., Sands, M.: The Feynman Lectures on Physics. Addison Wesley, Boston, Massachusetts (2005)
2. Seoane, L.F.: Optimality pressures toward lateralization of complex brain functions. Phys. Rev. X **13**, Paper 031028 (2023)
3. Thorne, K.S., Blandford, R.D.: Modern Classical Physics: Optics, Fluids, Plasmas, Elasticity, Relativity, and Statistical Physics. Princeton University Press, Princeton, New Jersey (2021)

GPSR Compliance

The European Union's (EU) General Product Safety Regulation (GPSR) is a set of rules that requires consumer products to be safe and our obligations to ensure this.

If you have any concerns about our products, you can contact us on ProductSafety@springernature.com

In case Publisher is established outside the EU, the EU authorized representative is:

Springer Nature Customer Service Center GmbH
Europaplatz 3
69115 Heidelberg, Germany

Batch number: 10767718

Printed by Printforce, the Netherlands